MW01633815

SAFETY AND THE BOTTOM LINE

Printed in U.S.A.

0-9656516-1-4

Published by
Institute Publishing, Incorporated
277 Main Street
Loganville, Georgia 30249

Library of Congress Cataloging in
Publication Data

Contents

Foreword

It is an honor to have this opportunity to write a foreword for the new book, *Safety and the Bottom Line*, authored by Frank E. Bird, Jr. and Ray J. Davies, world renowned experts in Safety and Loss Control. Since my personal definition of safety is "avoidance of loss," the philosophy presented here will show the reader how avoidance of loss will ultimately be recognized on the Bottom Line. Det Norske Veritas (DNV) has been in the business of safeguarding people, property and the environment for 132 years and our organization is committed to total safety management. This term extends the traditional objective of safety management beyond harm and injury to include environmental pollution, property damage, employee health and many other factors that affect the well-being of people and society.

DNV, the world's leading classification society, is one of the major certifying bodies for quality and environmental systems. In recent years, we have also become one of the major certifying bodies for quality and environmental systems. Industry recognizes the relevance of common standards in removing barriers to international trade. With our long experience in a number of loss control activities, it is only logical that one of our more recent goals is to also become the leading provider of total safety services.

Today's growing public concern and expectation means that our challenge is to put ever more effort into making the world safer, cleaner and fiscally acceptable. The world demands steps toward from the traditional "reactive" solution, toward a proactive, goal-setting management approach. *Safety and the Bottom Line* will give the reader the insight and ideas for setting the goals which will not only make the workplace a safer, cleaner and more productive environment, but will also show a measureable improvement in the Bottom Line.

Sven Ullring
CEO, Det Norske Veritas

Preface

A CEO and speaker at a major safety conference is quoted in Chapter 5, making a statement that professionals ought to know.

> "I think it is important for you to know what the typical CEO thinks about items that are important to him and how he feels about items that are important to the organization. Safety is not one of these items, it ranks very low on the totem pole. It ranks low because the CEO is interested in cost; he is interested in productivity; and he is interested in return on investment."

But, the most recent study of a significant sample of safety leaders reported that, it seems to make clear that the need for management support of safety continues to be one of the priority needs of safety professionals throughout U.S. industry. The study was reported in the article, 'Toward a National Agenda for Occupational Safety and Risk Control Management," by Professors Janice L. Thomas and Michael R. McDonald of the Virginia Commonwealth University, published in the *Professional Safety* magazine of the American Society of Safety Engineers, January 1994 issue.

This book presents a number of ideas concerning the benefits of safety to cost reduction and productivity, as well as its primary purpose in protection of our organizations' most valued resource.

The authors want to thank our colleague, a respected co-author of several of our books and master teacher, George

L. Germain for his splendid editorial help. Our special thanks to Paula Terry for her conscientious work as project assistant, and her excellent performance that enabled us to attain our production goals.

We wish to thank the following contributing authors for their chapters that have made such a valuable addition to this book:

> Stanley G. David, for his chapter on "A Quality Expert's View."
>
> George L. Germain, for his chapter on "Beyond Behaviorism to Holistic Motivation."
>
> Don J. Pedley, for his chapter on "Safety Pays."

In addition, many thanks to our friend and professional colleague, Dr. Douglas Clark, for his valued opinions.

Feel free to communicate with the authors for possible answers to questions that may arise or for additional references that may be available in your area of interest. With our wish that some thought herein may assist you in our continued endeavors to help others gain the commitment they need and seek for excellence in Safety.

Your colleagues in sharing,

Frank E. Bird, Jr.
Ray J. Davies

Dedication

This book is dedicated to Tricia Baker, Pat Bennett, Helen Bradford, Sandy Bratcher, Tina Cyphers, Teresa Dunn, Jenny Gates, Ruth Geiger, Linda Harper, Ruby Harrell, Penny Henderson, Joanne Laribee, Teresa Maddox, Diana Marshall, Dolores Rutledge, DeDe Skipper, Sheila Thompson, Iris Tyo, Liticia Weissinger, and Cindy Witcher for their continuing dedication and outstanding performance that enables colleagues to deliver the products and services used so widely around the world in the conservation of people, property and environment.

Chapter 1

SAFETY IMPROVES OPERATIONAL PERFORMANCE

"The power of the gulf stream will flow through an ordinary drinking straw if the straw is placed parallel to the flow of the stream."

Ralph Waldo Emerson

INTRODUCTION

This book is for those professionals who never cease their search for continuous improvement of all aspects of the management system, including protection of the most valuable resource, people. This search may be the reason so many leaders now emphasize integrating safety into all aspects of the management system. A widely accepted axiom used by leading experts on manufacturing excellence is that *our foremost concern in attaining quality must be with the quality of people.* W. Edwards Deming has made it perfectly clear that only after we have demonstrated our concern for, and taken care of, our "humanware" can we expect their concern and loyalty for development of the "hardware" and "software" aspects of our business.

We will clearly show how integrating safety management into every facet of the operational management system in your company will not only fulfill the advice of Mr. Deming and other quality gurus, but also can be the greatest "Return on Quality" (ROQ) measure that management can make today. One of the most popular speeches ever given was "Acres of Diamonds" by Russell H. Conwell. It was first given at a reunion of his Civil War comrades and followed by more than 5,000 deliveries. It earned him several million dollars which he used to found Temple University. The central message was very simple, "Riches of the value of diamonds are frequently right under your nose in your own back yard."

That is the message of this book. By simply integrating safety into every facet of the management system, operational performance, quality, waste control and the bottom line are given a giant "shot in the arm." Of course, the return on quality (ROQ) alone is enormous, as will be shown in Chapter 2. Doesn't it make sense that if we have common interests in things critical to safety, quality and productivity, we ought to be coordinating efforts to combine our strengths? Let's look at some critical items common to safety, quality and productivity:

- Attitude
- Skill
- Equipment
- Machinery

- Knowledge
- Physical fitness
- Materials
- Structures
- Environment
- Process

To gain the power portrayed in the earlier Ralph Waldo Emerson quotation, *our efforts to strengthen and protect the common items must flow together*. However, the majority of safety professionals fail to integrate or "flow" safety with quality and productivity. They go their separate ways, frequently duplicating other management work. As the "Acres of Diamonds," message communicates, you don't have to go outside and spend thousands, even millions, to make substantial improvement. "Riches of the value of diamonds are in your own backyard, right under your nose," in your present safety program. All that is required to gain them are a few relatively simple changes.

This chapter explains:

1. How to insure integration of safety into every aspect of the management system
2. How much the system is being used
3. What the goals are for a system approach
4. How integration improves operational performance
5. Other benefits of the system management approach
6. How executives using this approach perceive the results.

A bright young operating executive during one of our consultations with him and his subordinate colleagues said: "I now look at my career in terms of two periods: before and after I applied a system approach to the management of safety." That's exactly what this book does—shows how the application of safety management tools and techniques to the management system can directly and positively affect the attitude of your executives and the bottom line of your company.

THE SYSTEM APPROACH TO SAFETY MANAGEMENT

One of the biggest concerns of safety leaders today is that safety is not well integrated into management operations. There certainly are not many concerns expressed more frequently or given more keynote attention than this. Typical are the words of Jeffrey E. Castello, in a recent article, "Safety Management: The Winds Of Change," in the February, 1995 issue of *Professional Safety*, a publication of the American Society of Safety Engineers. "The safety manager's leadership and teamwork skills are a key to making companies more effective. The winds will require safety and health to be an integral part of any organization. The safety manager's skills are - and will continue to be - critical factors in helping others within the company realize their goals."

While Mr. Castello's words summarize beautifully what we believe to the marrow of our bones, the truth is that our safety programs are simply not well integrated, in spite of all of their potential benefits, into the management system. Management has insufficient reason to recognize the full potential value of safety to the organization. Might this also partially explain why management commitment to safety is not as complete as the profession would desire? We wonder, who really shares a major part of any fault for this apparent lack of commitment?

Perhaps no safety organization in the world states it better than the Industrial Accident Prevention Association (IAPA) of Ontario, Canada. An advertisement for a popular video, "Excellence in Safety," states that (1) safety must be an integral part of your business operation, (2) safety management improves the bottom line and (3) take a systems approach to the management of safety.

Logic and common sense confirm the many such statements by leaders today. The fact is that while thousands of companies around the world are successfully and profitably integrating safety through a systematic safety management approach, more and even most are not. Thus, the clarion call is to integrate, while the majority actually separate and even duplicate work. Then they wonder why executives and supervisors fail to give the support they believe the safety program justifies.

Maybe every now and then all of us in the safety profession should take just a few moments to look back at successes already accomplished to seek answers to major problems facing us today. The famous Madame Chang Kai-Shek is quoted from time to time as having said, "We live in the present, we dream in the future and we learn from the past."

Michael Le Boeuf, Ph.D., puts it a little differently in his book, *The Greatest Management Principle in the World*. He recalls the advice of one of his former professors many years ago, "When you can't understand a problem, go back to the basics and you'll start finding some answers. The greatest truths are too important to be new."

In the search for ways to assist in getting safety more deeply integrated into the management system, some insight may be gained by looking back to July 20, 1969 when Neil

Armstrong's first step on the lunar surface climaxed what could be considered the greatest safety achievement in history. A major contribution in making this great scientific achievement possible was the combination of safety and management skills that brought all successful life cycle stages of this fantastic process to pass. As we attempt to achieve success in our safety programming, need we be reminded that answers to our most critical problems can be found in past achievements?

This point is underlined boldly by a conversation with Jerome Lederer, who was Director, Manned Space Flight Safety, National Aeronautics and Space Administration, at the time of the moon landings. Mr. Lederer repeated a comment he must have made dozens of times in his many key note addresses.

> "With 5,600,000 parts and 1.5 million subsystems and assemblies, our strive for perfection in the Apollo 8 program produced a system that was 99.9% reliable. Unquestionably, it was the integration of safety into every aspect of the system that played a role in the accomplishment and you know, the 25 billion dollars to accomplish all this came back 7 dollars for every dollar spent and doesn't include all the dollars returned on new medical advances like the pacemaker."
>
> —Jerome Lederer

With the changes in programming methodology suggested in this chapter, a system approach to safety management can be implemented quite easily. It's just a matter of doing it... as some have already done with great success. All we have to do is make a few changes such as those described later in this chapter. Participants are likely to say these constitute not only a major step in the integration of safety but also a giant step toward building safety into the organizational culture and "the way things are done around here."

Reluctance to make the changes necessary is one of the biggest roadblocks to progress in every profession today, as it has been through the years. Over 30 years ago John Grimaldi, co-author of the popular book *Safety Management*, gave a speech that provoked questions about reluctance to change in safety management. The advice from this sage of safety management focuses on the biggest reason for many professionals not using a system approach. "An observer can easily conclude that the answer to accident problems may be given either in the existing safety recipes or in some which eventually will be devised. Under the circumstances, continued reliance on established corrective prescriptions is not surprising. Their acceptance is so strongly fixed and their application is so widespread that it seems almost impossible to de-emphasize reliance on them."

The authors of this book have participated in many conferences and have responded on dozens of questions related to a system approach. The biggest concern that professionals usually express for their failure to "flow with the gulf stream," is their fear of losing whatever level of management support they currently have. It's a paradox that the number one problem acknowledged by the majority is the need to gain more support. The encouraging message for those with such concerns, regardless of their cause is that the conversion is simple and easily acceptable for those whose support is necessary.

Zig Ziglar, one of America's most popular motivational speakers and authors points out...

> "You can get everything in life you want, if you help enough other people get what they want."

THE POWER OF HABIT

You may know me.

I'm your constant companion

I'm your greatest helper: I'm your heaviest burden.

I will push you onward or drag you down to failure.

I am at your command.

Half the tasks you do might as well be turned over to me. I'm able to do them quickly, and I'm able to do them the same every time if that's what you want.

I'm easily managed, all you've got to do is be firm with me.

Show me exactly how you want it done; after a few lessons I'll do it automatically.

I am the servant of all great men and women; of course, servant of the failures as well.

I've made all the great individuals who have ever been great.

And I've made all the failures, too.

But I work with all the precision of a marvelous computer with the intelligence of a human being.

You may run me for profit, or you may run me to ruin; it makes no difference to me.

Take me. Be easy with me and I will destroy you.

Be firm with me and I'll put the world at your feet.

Who am I?

I'm Habit!

Anonymous

We put it just a little differently in one of our books, *Practical Loss Control Leadership*, referring to an important principle of program implementation we call the Principle of Mutual Interest. "Program projects and ideas are best sold when they bridge the wants and desires of both parties." As a matter of fact, people who are best at selling programs or ideas are those who clearly establish a bridge or connection of values between what the company wants and what they want.

The switch to a system approach to safety management is so easy because of its common objectives with those of operating management. The conversion can be made step by step, program activity by program activity, without a hitch.

This truth is also supported by Richard C. Grote, an international speaker and authority on management development and organizational behavior. In one of his many published articles, "Bringing About a System Change," he points out that, "Most systems fail not because of inherent flaws, but because nobody's making sure they are maintained and linked to all other programs or systems." Again, isn't this exactly why most safety programs fail to get the management commitment they seek?

Let's look at the goals of a system approach to safety management to see how little they differ from a purely injury or illness related endeavor.

- Goal No. 1. Clearly identify program elements/activities common to successful programs and establish standards/requirements for performance at each organizational level.
- Goal No. 2. Establish an objective system for measuring and evaluating work unit and individual conformance with program standards/requirements,

commending positive performance and constructively correcting substandard performance.

- Goal No. 3. Identify and motivate the specific behaviors, of each executive and the management team that strengthen personal commitment to the safety system and demonstrate it to all employees.
- Goal No. 4. Establish an effective review system to identify and control potential causes of loss at the conception, design, implementation and disposal stages of engineering activities.
- Goal No. 5. Identify and apply specific communication and motivation activities which ensure that all employees have and use the required knowledge and skills to do their work "the right way."
- Goal No. 6. Systematically identify the critical few jobs/tasks that have been or could be involved in 80% of the behaviors related to major or catastrophic loss events to people, property and/or process, and develop effective procedures and/or practices to guide loss-free performance.
- Goal No. 7. Systematically identify and establish an inspection program for the critical few parts/items that if worn, damaged or not operating properly, have been or could be involved in 80% of people, property or process loss events.
- Goal No. 8. Systematically identify all substances and materials that could be involved with major or catastrophic loss if not properly manufactured, stored, transported, distributed, utilized and disposed of.
- Goal No. 9. Establish a comprehensive emergency preparedness plan and take action to ensure its effective operation.

- Goal No. 10. Systematically identify the costs involved with the critical few items of injury, damage and waste that have historically related to 80% of loss, and apply problem-solving actions for prompt profit improvement. (This provides the motivation for upper management to give you the commitment necessary to implement and achieve the first nine goals).

A typical comprehensive world-class safety program being used by thousands of companies today would include the following elements:

- Leadership and Administration
- Leadership Training
- Planned Inspections and Maintenance
- Critical Task Analysis and Procedures
- Accident/Incident Investigation
- Task Observation
- Emergency Preparedness
- Rules and Work Permits
- Accident/Incident Analysis
- Knowledge and Skill Training
- Personal Protective Equipment
- Health and Hygiene Control
- System Evaluation
- Engineering and Change Management
- Personal Communications
- Group Communications
- General Promotion
- Hiring and Placement
- Materials and Services Management
- Off-the-Job Safety

While this chapter will not provide a step-by-step map for integrating safety into the management system, it will provide enough information to kindle the spark of motivation that will ignite the desire to do so.

The chart below indicates a few of the paradigm shifts that a system (or holistic) approach to safety requires.

Traditional Approaches	System Paradigm Shifts
Unsafe acts and conditions	Substandard acts and conditions
Job safety procedures	**Standard job/task procedures**
Job safety observations	**Planned job/task observation** (including safety, quality and production)
Inspections (for hazards)	**Planned inspections** (potential problems, equipment deficiencies, substandard employee actions, significant changes)
Hazard report form	**Condition report form**
Unsafe act correction	**Commendations and correction**
Frequency and severity rates (primary measurements)	**Individual and system performance** (primary measurement)
Job safety instruction	**Proper job/task instruction**
Accident investigation report form	**Investigation report** (Actual and potential loss events)

Those few examples are typical of paradigm shifts necessary for a system approach to safety management. If the necessary paradigm shifts are made, here are some of the specific ways a few of the elements of a system approach to safety would improve operational performance and contribute to the bottom line.

Management Training

Most readers probably will agree that a formal safety training session, including all elements of the program, should be held every few years. Of course, loss experience will tend to influence the amount of emphasis or time devoted to individual elements. Successful professionals will also tend to agree that such training is not enough. Over and over again in articles, posters and meetings, hundreds of minuscule training reinforcements are presented as reminders and motivators for doing the job properly. This demonstrates a significant difference between typical programs with regular training activities and the continuous training activity done in a successful safety and health system.

In many of his speeches Knute Rockne, the famous football coach at Notre Dame, described his training method that was really typical of many other great coaches.

> "At Notre Dame, we have about 300 lads, both varsity and newcomers. I keep them practicing fundamentals over, and over, and over again until the various fundamentals become as natural and subconscious as breathing. Then, in the game, they don't have to stop and wonder what to do next when the time comes for quick action."

Whether it's emphasizing and re-emphasizing the way to give proper job instruction (including safety), or whatever safety program activity is the point of emphasis — if it's system related (doing the job right), leaders will not only listen more but do it more. They will put their greatest effort where they know top management wants them to put it.

Safety research described later in this chapter points out clearly that the practice of separating safety from the right way to do the job hasn't achieved our mission of preventing accidents to the level we desire. Neither has it gained real management commitment. These goals are more likely to be reached through the system approach.

The frequent emphasis of a system approach that gets things done properly (including safely) will pay off in improved management performance just as doing it right, over and over in sports wins games.

Planned Inspections

The identification of safety, health and environmental hazards alone make every minute spent on the regular plant inspection justified. However, operating management will be more highly motivated to support this important activity when the inspections also identify:

- important potential quality or production problems
- significant equipment deficiencies
- employee actions worthy of commendation or in need of correction
- significant items of waste.

We know that the thousands of companies utilizing a holistic approach give safety, health and environment special attention on their planned inspection tours. We also know that it adds power and importance to this inspection activity when identifying levels of compliance with other subjects high on management's motivational scale is part of the exercise. Again, isn't it logical that it all spells total system improvement?

Task Analysis and Procedures

Task analysis and procedures is a key element of most safety and health programs that can contribute more equally to quality, productivity and costs, as well as safety. It's interesting that a number of companies using a system approach to safety have reported to the authors that on the strength of their work on this program element alone, the ISO 9000 Certification was 70-80% easier to attain. Identification of all of an organization's critical tasks that could cause a major or catastrophic accident, quality or productivity loss if not done properly are identified in the task analysis process. More frequently than not, this is the first time that the vast majority of organizations have become aware of their critical tasks in a systematic and inclusive manner. *Figures 1-1 and 1-2* illustrate improved operational performance resulting from this element, under the system approach.

It is not uncommon for an organization to have 50 or more occupations, including 200 or more critical tasks that, if not performed properly, could result in major loss. Losses frequently involve injury as well as job inefficiency. It is also common for one or more task improvements to be made each time a task is analyzed or a task procedure changed. While the cost reduction in *Figure 1-1* is relatively small, if only one such improvement were made every two years on 200 critical tasks, it could represent as much as $800,000 in cost improvement. This certainly points out the importance of keeping records of all improvements made from doing task analysis and procedures. As most professionals know, an improvement check of job/task steps is done with the employee who performs the work before the procedure for doing the task is developed. The potential of finding a better or more efficient way to do a task is always one of the objectives. Interestingly enough, improvements found during this facet of the task analysis frequently pay for all time spent on this program element.

Continued task improvement potential is also built into the updates of procedures. For example, a critical task procedure is updated:

- whenever a change is made to anything related to doing the task.
- whenever a serious or major problem (safety, quality or productivity) occurs related to the task.
- at not less than a predetermined update rate (such as yearly), based on task criticality.

It's not difficult to see how this aspect alone, which supports continuous improvement, has great appeal to

Task Analysis Improvement Suggestion Report

LOCATION	PLANT/DIVISION	DEPARTMENT	OCCUPATION	TASK
	Concentrator	Grinding	Grinder Operator	Start Particle Size Monitor
	DATE	SUBMITTED BY	ANNUAL COST SAVINGS	APPROVED BY
	5/18/19–	S. Bentley	$4,017.00	K. Brown

DESCRIPTION OF CHANGES AND BENEFITS

CLEARLY DESCRIBE ALL CHANGES/EFFICIENCY IMPROVEMENTS MADE OR THAT WILL BE MADE AS A RESULT OF THE IMPROVEMENT CHECK.

Previously the sensor pump had to be removed from the monitor for clearing when it was clogged. This had to be done an average of 4 times a day and took about 10 minutes to remove, clear and replace. Also, because the operator would occasionally try to remove it without shutting down, there was a serious potential for hand injury. Manipulating the wrench in such tight quarters frequently led to wrench slippage and a bruised hand.

CLEARLY DESCRIBE ALL BENEFITS (DIRECT AND INDIRECT) THAT HAVE OCCURRED OR WILL OCCUR BECAUSE OF THESE CHANGES.

The new procedure uses a special fitting which allows the pump to be cleared while still mounted. This only takes about one minute and the operator's hand is never in danger. Also, it eliminates the annoyance and hand bruises from using the wrench for disassembly and assembly. The equipment should still be shut down to avoid damage but if for any reason it is not there is no danger of injury.

COMPUTATION OF SAVINGS

CLEARLY DESCRIBE METHOD USED TO COMPUTE AND MEASURE COSTS, BEING SURE TO INCLUDE IMPLEMENTATION COST.

Ten minutes previously x 4 times daily	=	40 minutes
One minute now x 4 times daily	=	4 minutes
Time savings (daily)	=	36 minutes
Time savings yearly–350 days x 36 minutes	=	12,600 minutes
	=	210 Hours

Hours per year	=	210
Labor costs	=	19.50
Labor savings per year	=	$4,095
Less cost of fabricating new tool		78
Annual savings first year		$4,017

Special Note: With only one small task improvement per critical task made every two years on 200 critical tasks, cost improvement could average well over $400,000 annually.

Figure 1-1

Plant/Division	Task Analyzed	TASK ANALYSIS WORKSHEET	W. Livingstone / Eng. / 5/18/19–
Concentrator	Operate Particle Size Monitor		Signature / Function / Date
Department	Date Completed		R. Hagan / Supv. / 5/16/19–
Grinding	5/14/19–		Signature / Function / Date
Occupation	Completed By		R. Swinehart / Supv. / 5/23/19–
Grinding Operator	E. Bromley		Signature / Function / Date

No.	Significant Steps or Critical Activities	Loss Exposures (Safety-Quality-Production)	Efficiency Check: Yes	Efficiency Check: No	Recommended Controls
1	Inspect Equipment	Equipment damage, poor readings	✓		Follow checklist
2	Look for sand buildup in cyclone box	Equipment failure will force coarse	✓		Clean any accumulation
		particles into system			
3	Hose out air eliminator	Damage to air eliminator from sand	✓		Clean any accumulation
4	Close drain valve	Water spills, electric shock	✓		Close drain valve
5	Open fresh water valve	Inadequate flow of water will cause	✓		Clear water filter
		plug-ups			
6	Clear sensor pump	Hand injury from wrench slipping	✓		Shut down before cleaning, use
					special cleaning tool
7	Ensure sensor is in place	Poor reading	✓		Replace if missing
8	Put safety system on automatic	Damage from low water pressure	✓		Put on automatic
9	Ensure that sample screens are clear	Poor readings	✓		Visually inspect. Clean any
	in cyclone overflow box				accumulation
10	Start Particle Size Monitor		✓		Pull stop/start button
11	Ensure adequate water pressure	Automatic shut down of system	✓		Ensure at least 60 psi
12	Allow tank to fill	Overflow, clogs, system shutdown			Listen for safety water to cycle off
					and on
13	Place sample screen in slurry	Inadequate vacuum	✓		Look for normal vacuum of 17+ inches
14	Adjust sample intake flow	Overflow or inadequate flow of slurry.	✓		Allow 5 minutes to stabilize
		slips and falls to surface			

Figure 1-2

operating managers. Obviously, the benefits from improving operating performance are great. Safety, health and environment also benefit from this concentration on the right way to do the job.

It's not necessary to go through all elements of the typical successful program to make the point that operating performance is improved. Ask any athletic coach how much running the plays and rerunning them over and over contributes to winning games and his/her answer, not surprisingly, will be on the high side. Perhaps it is the ability of athletic coaches to improve performance that has motivated so much emphasis on adopting *COACHING* techniques into management practices. Unfortunately, most organizations do not but in successful safety management systems it is done with religious zeal.

Let's look at several other benefits from the system approach to safety.

SYSTEM APPROACH BENEFITS

Increased Management Commitment

Perhaps the biggest benefit of a system or holistic approach is the increased commitment that continually grows, due to the perception of management at all levels that the safety program contributes to the concerns that are of high priority to everyone. Comments the authors have heard a number of times from various levels of operating management in companies using this approach are similar to the following statement.

> "Everyone knows the safety program strengthens every aspect of our management system."

The critical need for management commitment and support is well recognized as one of the biggest needs in safety management today.

Research Results Show the Need

The results of an extensive study appeared in the article "Toward a National Research Agenda for Occupational Safety and Risk Control Management," by Professors Janice L. Thomas and Michael R. McDonald, of the Virginia Commonwealth University in Richmond, Va., in the January, 1994, issue of *Professional Safety*. The study involved a significant number of American Society of Safety Engineer's members who are responsible for managing safety and health programs within large organizations, as well as other professionals representing risk-control programs, research scientists and regulatory administrators. The study established "The 27 Most Important Research Topics/ Needs," *(Figure 1-3)*.

One of the four top research needs today in occupational safety and risk control management, follows:

> "Assess methods for developing management commitment to safety/risk-control."

The rating to classify the importance was based on a ranking of low (1) to high (5) in terms of importance. Note that several of the 27 needs relate to commitment. Experience and common sense tell us that top management's commitment to safety grows with their perception of its value to operational performance.

One reason for lack of management commitment may be that so few managers are aware of the contributions to be made by a systematic approach to safety.

27 Most Important Research Topics/Needs

Rating	Research Topics/Needs
4.13	Evaluate and project the impact of the aging workforce on the occupational safety/health loss experience.
4.13	Develop and validate a hazardous products identification system appropriate for use among populations with limited reading ability.
4.00	Assess methods for developing management commitment to safety/risk control.
4.00	Compare and evaluate traditional criteria used to measure risk control effectiveness (i.e., incidence rates, OSHA citations and penalties, experience modification rates, medical and workers' compensation losses, etc.).
3.88	Study and profile the accident experience of the phantom workforce (i.e., temporary employees).
3.75	Evaluate the most effective methods to achieve the "proactive" incorporation of risk control principles into the early design of facilities, processes and management systems.
3.75	Study and profile the culture within organizations evaluated to have achieved and maintained high levels of risk control effectiveness.
3.75	Evaluate and recommend a model accident/mishap "cost accounting" framework or system that may be used by safety managers to document loss problems and to justify and evaluate safety programs.
3.75	Analyze and evaluate factors associated with rising incidence of cumulative trauma and psychological stress workers' compensation claims.
3.75	Evaluate links between cumulative trauma disorders and predisposing physical conditions.
3.75	Develop methods to encourage and incorporate use of safety into CAD systems targeted for use by professional engineers and facility designers.
3.63	Assess the utility of various evaluation models and techniques for determining safety program effectiveness.
3.63	Formulate and evaluate marketing techniques that will attract high school and college students to the safety and risk control profession.

Figure 1-3

3.63	Evaluate the relevance of ASSE's model safety curriculum guidelines.
3.50	Assess problems and conditions that block the full incorporation of safety into the organizational management framework.
3.50	Develop a multifaceted index for use in measuring and evaluating organizational safety and risk control performance (to include organizational climate, various loss rates, behavior survey data, etc.).
3.50	Inventory and analyze the current state-of-the-practice regarding emergency planning and response within occupational/industrial settings.
3.50	Develop/validate equations (similar to NIOSH lifting equations) useful for assessing risks associated with standing, sitting, push, pull and reach activities.
3.50	Inventory and evaluate problems associated with contractual workers as they relate to fixed site safety problems.
3.38	Study alternative methods for developing employee commitment to safety and risk control programs.
3.38	Study the factors that contribute to senior level management commitment to safety and risk control.
3.38	Evaluate methods for increasing supervisory commitment and accountability for a safe work environment.
3.38	Conduct a national study of safety incentive programs and their relative effectiveness in instilling and maintaining desired behaviors.
3.38	Design a detailed conceptual framework to guide manual and automated hazard identification and risk analyses.
3.38	Develop a conceptually complete, multi-tiered accident data reporting, collection and analysis model that integrates and responds to various stages and levels of organizational need and sophistication (i.e., that can advance in scope and depth with advancement of organization).
3.38	Conduct a national assessment of academic programs in safety and evaluate present and future sufficiencies of safety/risk control graduates.
3.38	Survey the availability and success of safety and health courses within college and university engineering curricula. Propose strategies for further development and integration.

Figure 1-3 continued

If a safety and health program was as important as industrial human relations experts wrote that it was; and as important to the control of such things as morale, presenteeism, labor relations, job behavior and performance (as well as control of injury and illness), then it was logical that safety should have an important place in the curriculum of every business school educating future industrial leaders. Questionnaires were sent to over 2,000 schools of business to determine if they included safety in the education of future leaders, only nine responded affirmatively. As Bill Pope said in his fine book, *Managing For Performance Perfection*, "the conclusion was that less than 1% of more than 1.6 million management graduates, over a ten-year period, received any significant exposure to safety education. They entered the practical business world without any appreciation of the industrial safety problem."

This study was given wide publicity and developed into a project to encourage business schools to incorporate safety into their study programs. It acquired the name "Minerva" from the Roman goddess of wisdom and became known as the Minerva Project. One of the many groups expressing interest in the Minerva Project was the National Institute for Occupational Safety and Health (NIOSH), a division of the U.S. Department of Labor. After a series of discussions, it seemed logical that the project would receive the greatest chance of success if sponsored by NIOSH. They assigned it with a grant to the College of Business Administration, Xavier University, one of the few schools already including safety in its business education. Now it is known as The Minerva Education Institute, sponsored by NIOSH under the aegis of Xavier University. To educate tomorrow's leaders in safety, health and environment seemed another step in the direction of that elusive management commitment. It's quite interesting that, 15 years after the inauguration of the project, a check by the authors with faculty at Xavier revealed less than 3 additional schools of business administration had incorporated safety into their curriculum for future industrial leaders. Further conversation revealed that very few current management textbooks currently in use include safety in their subject content.

The purpose of all this chapter's groundwork is not just to reaffirm the obvious point that gaining management commitment remains the number one objective to advance industrial safety. In addition, the following hypothesis is central to the purpose of this book: Time may be better invested if we examine the reasons for our failure to gain the support needed to cope with safety, health and environmental challenges that could threaten the continuation of our business existence as powerfully as worldwide competition does.

The positive feedback that will come to key executives from others in the organization, once safety is integrated, will also contribute greatly to their positive perception. On the other hand, safety as an adjunct activity requiring extra, or duplicated work will have the understandably negative effects on commitment reflected in a significant number of the 27 research needs listed previously.

Environmental Risk Control

Thousands of major companies around the world that use a systematic safety management approach have taken programs designed with a holistic approach to accident control in order to maximize the efficiency of their people. In these organizations, a supervisor/leader will report an environmental or health hazard when making a safety inspection, doing a job analysis or planned observation. In tune with today's work improvement philosophy, we must minimize the duplication of efforts necessary to ensure that the work is being done the right way.

The President of the National Safety Council was a featured speaker at the 38th Annual Western Safety Congress in Los Angeles. He made it crystal clear that the big barrier to success in process safety was the number of audits required that barrage operating management. Mr. Scannel stated, "There is a big need for one audit that includes safety, health and environmental control. It is ridiculous that key personnel must spend hours, even days of highly valued time answering the same questions about activities common to these three important areas."

Let's again look at elements of a modern comprehensive safety management system.

- Leadership and Administration
- Leadership Training
- Planned Inspections and Maintenance
- Critical Task Analysis and Procedures
- Accident/Incident Investigation
- Task Observation
- Emergency Preparedness
- Rules and Work Permits
- Accident/Incident Analysis
- Knowledge and Skill Training
- Personal Protective Equipment
- Health and Hygiene Control
- System Evaluation
- Engineering and Change Management
- Personal Communications
- Group Communications
- General Promotion
- Hiring and Placement
- Materials and Services Management
- Off-the-Job Safety

It is easy to see how the vast majority can apply to safety, health and environment. Of course, there is still the need for a few specific health or environmental control elements, such as environmental controls dealing with specific hazards and issues, specific performance monitoring and assessment, relations with external parties and product stewardship.

With all the unnecessary wasted time and energy some leaders or executives require of their management, it is not difficult to understand why their commitment is not adequate. A well-designed program, in tune with world class standards, will be able to more promptly and effectively meet the ever increasing environmental risk control concerns.

Prevention of Errors and Catastrophes

Another major benefit of the system approach is prevention of the errors that cause major and catastrophic events. When safety is an integrated part of the right way to do work and no longer something separate and extra, management compliance with all related standards improves. Safety benefits from the attention people give to those things they know management wants them to attend to. In effect, safety is no longer a second-class citizen, but part of the mainstream of management activity. The big contributor to the control of major loss events is the reduced error rate that comes with compliance to all standards.

Since safety receives more attention as an integrated part of the system, the errors that cause all losses, including accidents, are minimized. With this improvement in compliance to all work standards, the chance of those deviations that could cause major loss or catastrophe is likewise minimized. To restate and reinforce this major benefit that comes from integration: with increased compliance to safety standards as part of "the right way" to do things, thus receiving more management support, there is less chance that major loss events, including catastrophes, will occur.

Many articles have been written on disaster prevention or lessons learned from past major events. Contributions of significant published data and studies on the human phenomenon of error involved with catastrophes have appeared in many popular professional journals in recent years. After reading and analyzing the literature on this vital area of concern, we feel that much of what has been said can be summarized in five basic principles.

Principles of Catastrophe Control

1. Principle of Systematic Control

   The higher the risk, the more systematic the controls should be to achieve required program objectives or goals.

2. Principle of Situation Control

   The higher the risk, the more effort should be directed at building a safety and quality prone environment/situation.

3. Principle of Stress Control

   The more critical the work, the more that effort should be directed at recognizing, evaluating and controlling causes of stress in the total organizational environment.

4. Principle of Checks and Balances

   The more critical the work, the more effort should be directed at building checks and balances into activities to ensure that they will be done the right way.

5. Principle of Forcing Mechanisms

   The more critical the work, the more effort should be directed at developing and applying management

> techniques, motivational processes, and mechanical devices that provide the forcing mechanisms which increase the likelihood of desired behavior and error free work.

Since all of these principles are built into the goals of the integrated approach, the increased commitment that comes with the right way of doing things gives much more needed attention to the prevention of errors that cause undesired conditions and events.

Jerome Lederer, referred to earlier, said this in one of his speeches following the moon landing:

> "This nation was built on risk, personal risk in tackling the wilderness, financial risk in business, risks in exploring the scientific unknown, enormous engineering risks, management risks. We shall continue to take risks of greater magnitude than in the past. But, the consequences of failure are becoming less permissible. The political, social, as well as economic and personal risks that now accompany our ventures can have enormous repercussions when failure occurs."

At the time of these words, Mr. Lederer would never, in his wildest imagination, have guessed the number of catastrophic events that would occur in the 30 years following his prophetic words of wisdom. There have been over 100 major fires or explosions in the hydrocarbon-chemical industries alone, averaging over 63 million dollars in property damage in addition to the thousands of lives lost and many more thousands injured.

These tremendous losses are only part of the industrial disaster scene. Bhopal, one of the worst industrial/environmental accidents of the 20th century, occurred in December, 1984, killing 2,500. The worst accident in nuclear power history occurred in April, 1986, at the Chernobyl nuclear power plant, leaving 31 people dead in the immediate aftermath but spreading significant quantities of radioactive materials over much of Europe. As a result of this radiation release into the atmosphere, tens of thousands of cancer deaths (as well as increased rates of birth defects) have occurred, with others expected for decades ahead. Without continuing to address the many other disasters that have occurred, it has become obvious to many outspoken safety leaders that the bleak history of catastrophic events points out clearly that the experience of disaster alone does not gain the continued commitment required to prevent them.

Greater control of catastrophes then, is the second big benefit that comes with the improved compliance to standards for doing the job the right way, which includes the safe way. Thus, the second value of an integrated approach to safety management.

Expanding Management Vision from Compliance to Excellence

The third benefit of the system approach is the recognition that there is much more to safety than compliance with the law. Top management's fixation with legislative compliance to avoid fines and imprisonment is the subject of numerous articles in professional journals. They reflect the concern of safety and health leaders for what is obviously an out-of-control misdirection. This is not to deny that compliance should be one of the important goals of a modern safety management system. The big concern is that too many executives apparently see this as the primary, even exclusive, objective. As a result, other critical safety goals do not get the support they justify.

While we are aware of other published concerns about this issue, none makes the point more forthrightly than an editorial in the April, 1995, issue of *Professional Safety* by Margaret M. Carroll, the outgoing President of the American Society of Safety Engineers. The title of the editorial was, "Is Our Role Changing?" The subtitle, "A strict compliance mentality turns us away from the real issues," gives partial insight to the straightforward cautions that follow. Perhaps none more clearly defines this misdirection than this terse statement by one of America's most respected safety leaders, "A disturbing attitude is spreading among management in corporations of all sizes and natures. At a time when growing numbers in the safety profession are formally educated in the discipline and in management techniques, the 'keep me out of jail' approach to occupational safety and health is negating their efforts."

Ms. Carroll points out that meeting the overwhelming number of these requirements (it has been estimated by various speakers that 50,000 now exist) may prevent them from allocating resources to move their programs from compliance to a position of safety excellence, "*...they opt for the bare minimum and direct the safety professional to ensure compliance.*"

Within the ranks of safety leaders compliance to regulations is alleged by many highly respected members to have little effect on the significant problems involved with injury and illness causation. In fact, several reported studies *within their corporations have indicated that less than 10% of their losses would be prevented by regulatory compliance.* At the same time, the attention required to maintain compliance with the vast number of regulations mentioned earlier saps budgetary resources completely.

Management's insistence on compliance is, on one hand, a

dream come true and certainly an action that would normally be applauded; but to accomplish this, while ignoring other aspects of injury/illness control is ignored, is misdirected commitment.

Management fixation with compliance to OSHA, EPA, and other laws is overcome as the much broader additional values of safety are more clearly recognized, and commitment is thereby increased. While the system or holistic approach is not employed by the majority of companies, we are elated that some of the world's largest and best known, have made the necessary paradigm shifts to expand their protection efforts more broadly.

Executive response to the system or holistic approach can be seen in the responses of 173 senior executives using this approach who participated in a research project to determine its effectiveness on organizational performance (see *Figure 1-4*). The project was motivated by the continuous requests at safety conferences for reference to any unbiased study of the value executives feel a good safety program has on their overall management system. When the study was proposed, we quickly volunteered to provide audit data on a large group of organizations using a system or holistic approach that we had introduced to them. The study was directed by Dr. Larry Gaunt, Professor of Risk Management and Insurance at Georgia State University. Dr. Gaunt and his colleagues prepared a questionnaire for rating the impact of their safety management systems on 29 organizational factors in addition to five occupational injury and illness factors. The 34 factors are listed in *Figure 1-4*. Executives from 173 organizations agreed to participate, using the Likert scale, shown on the following page, to indicate their perceptions of the effect of their safety management systems on each of the 34 factors.

LIKERT SCALE	
Scale	Score
Strong positive effect	7
Moderate positive effect	6
Slight positive effect	5
No effect	4
Slight negative effect	3
Moderate negative effect	2
Strong negative effect	1
N/A	0

A score above four on this scale shows a positive effect and a score below four shows a negative effect.

There was great consistency among all those participating that the five organizational factors most positively affected by the safety management system were:

1. Housekeeping
2. Management Practices
3. Work Methods
4. Worker Participation
5. Supervisory Skills.

The organizations were stratified by level of system implementation as verified by objective audits, with program audit level five being the highest. The results are shown below:

ORGANIZATIONAL PERFORMANCE RATINGS AT PROGRAM LEVEL 4		
Factor	Mean	Mode
Housekeeping	6.47	7
Work Methods	6.27	7
Management Practices	6.00	6
Worker Participation	5.80	7
Supervisory Skills	5.73	6

Organizational Performance Factors

Factor #	Factor	Factor #	Factor
1	Inventory Control	18	Space Utilization
2	Product Quality	19	Material Waste
3	Work Methods	20	Labor Turnover
4	Worker Participation	21	Absenteeism
5	Labor-Management Relations	22	Worker Morale
6	Public Relations	23	Public Liability Hazards
7	Customer Relations	24	Employee Suggestions
8	Regulatory Compliance	25	Job Pride
9	Operating Costs	26	Work Process Efficiency
10	Housekeeping	27	Product Rejection Rate
11	Worker Satisfaction	28	Profitability
12	Employee Grievances	29	Employee Commitment
13	Management Practices	30	Occupational Injury and Illness Cases
14	Productivity	31	First Aid Cases
15	Employee Work Behavior	32	Lost Workday Cases
16	Supervisory Skills	33	Days Away / Restricted
17	Property Damage	34	Summary

Figure 1-4

CONCLUSION

Integrating safety into every facet of the management system, gives you a giant boost to operational performance. Other chapters will show how this also affects quality, waste control and the bottom line.

- Don't separate. Don't duplicate. Do integrate.
- Conduct continuous training; become a learning organization.
- Inspect for safety, health and environmental hazards; potential equipment deficiencies; waste; and employee actions in need of correction and/or worthy of commendation.
- Broaden task analyses, task procedures and other safety activities from "the safe way" to "the right way."
- Effect continuous cost improvement through continuous task improvement.
- Apply a holistic system to the prevention of errors and catastrophes.
- Expand the management vision from compliance to excellence.

Discover the "Acres of Diamonds" in your own backyard. The big gem of "Improving Operating Performance" was discussed in this chapter. The gems of Quality Proneness and Waste Control Improvement from your safety program are discussed in other chapters. Mine and profit from all these beneficial management gems.

Selected References

Castello, Jeffrey E. "The Winds of Change." *Professional Safety*. Illinois: ASSE, 1995.

Conway, William E. *The Quality Secret, The Right Way to Manage*. Nashua, NH: Conway Quality Inc., 1992.

Garrett, Larry D. *The Effect of the International Safety Rating System on Organization Performance.* Center for Risk Management and Insurance Research. Atlanta: Georgia State University, 1990.

Gaunt, Dr. Larry D. The Center for Risk Management and Insurance Research, Atlanta, Georgia State University.

Gethoro, Howard S. and Shelly J. Gillow. *The Deming Guide to Quality and Competitive Position*. Englewood, NJ: Prentice Hall, Inc., 1987.

Grimaldi, John V. and Rollin H. Simonds. *Safety Management*. Boston, MA: Irwin, 1989.

IAPA. *Excellence in Safety*. Video. Ontario, Canada, 1992.

Le Boeuf, Michael. *The Greatest Management Principle in the World*. New York: G. P. Putmans' Sons, 1985.

Marshall, Gilbert. *Safety Engineering*. Des Plains, IL: ASSE, 1994.

Plank, Thomas W. and Kevin T. Fern. "Reevaluating Occupational Safety Priorities." *Professional Safety*. Des Plains, IL: ASSE, 1993

Pope, William C. *Managing for Performance Perfection—The Changing Emphasis*. Weaverville, NC: Bonnie Bray Publications, 1990.

Rust, Roland T., Anthony J. Zahorik, Timothy L. Keiningham. *Return on Quality*. Chicago, IL: Probus Publishing Company, 1994.

Smith, Roger. *Catastrophes and Disasters*. New York: Chambers, 1992.

Thomas, Janice L. and Michael R. McDonald. "Toward a National Agenda for Occupational Safety and Risk Control Management." *Professional Safety*. Des Plains, IL: ASSE, 1994.

Chapter 2

SAFETY DEVELOPS A QUALITY PRONE ENVIRONMENT

"The appearance of a place is a self-portrait of the person who works there. Autograph this image of your work with excellence."

Anonymous

Leading experts have said that having the work environment in a state of quality proneness is the first step in achieving excellence in quality products. Perhaps no one has expressed this more pointedly than Alan Swain, now retired and formerly with the Sandia Corporation in Albuquerque, New Mexico. Dr. Swain is internationally respected for his pioneering human reliability work in the nuclear weapons field and is frequently referred to as the father of human reliability analysis. One of his frequently used comments is, "One of the best ways to keep workers doing their job properly is to create a quality prone environment in which to work." This is both a logical and common sense statement. For example, who could possibly contest the logic of having all necessary tools, equipment, raw materials, etc. in their proper place at all times during the work process in order to turn out quality products consistently?

Dr. Jorma Saari, is a noted safety researcher from Finland, who recently completed a significant study on housekeeping. He states it this way, "When there is order, it is easy, almost automatic, to find whatever is needed." Likewise, doesn't it make a great deal of sense that the workplace should be clear of all unnecessary items? And, isn't it logical that storage of finished products should reflect an orderly arrangement best suited to finding specific items and checking inventory level, as well as impressing customers during their periodic visits?

If your answer to these questions directed at fundamental quality management basics is an anticipated "yes," get ready for a big shock. In spite of the importance of these basics, many if not most executives do not have an organized program to maintain and continuously improve the state of order/housekeeping. It has been said that the relatively new method of objectively rating/measuring "order," to effectively and accurately know the housekeeping level, is one of the greatest "Return on Quality" actions that any management could possibly take for the small investment of time required. This will be discussed later.

Obviously, when such performance areas as work methods,

management practices and supervisory skills are perceived to be positively affected in a significant way by the safety program, as shown by the research in Chapter 1, quality can benefit. Remember the study by Dr. Gaunt naming housekeeping as the factor perceived to be the highest benefactor of safety at all levels of program. We feel very strongly that management in general has under-evaluated the benefit to any manufacturing organization from a clean, orderly workplace. Housekeeping not only creates a quality prone environment but also a quality prone attitude. This may be one of the best kept quality secrets today.

The objectives of this chapter are to:

1. Highlight the benefits of order for quality
2. Show how order stimulates pride in performance
3. Explain the concept of order
4. Provide practical implementation steps.

A number of the comments in this chapter are from the book *Profits Are In Order*, available from the Det Norske Veritas in Loganville, Georgia. Because of our deep belief that housekeeping makes an enormous contribution to the management system, including a strong positive effect on employee attitude, we felt that information currently available on this seemingly mundane but vitally important subject ought to be suitably assembled and recorded in one volume. The book has produced such unexpected reactions from executives both here and abroad, that a reprint with a postscript and answers to some of the inquiries has been published.

Let's examine a few relevant comments from *Profits Are In Order* that motivated us to include this postscript.

POSTSCRIPT
to
Profits Are In Order

Since the initial printing of this book, I have had a surprising number of communications from safety/operating managers of companies here and abroad. The reactions shared were of such interest, particularly those of operating managers, that the idea of adding this postscript evolved.

As an example, I was not fully aware of the number of managers searching for additional ways to add to their continued improvement programs and other aspects of business without making further expenditures in programs whose return might not fulfill expectations.

As one manager put it, "I thought I'd read a small portion of your book on order, out of courtesy to my safety director. He had channeled a copy to me highlighting the linkage between housekeeping and the amount of defective products produced. After reading a paragraph or two, I couldn't put the thing down. I've encouraged good housekeeping for years, but in all honesty, I saw it only as something accepted for managers to do. I didn't really see it as a natural, inexpensive and major adjunct to our improvement program. Perhaps I'd be more truthful if I admitted I didn't see it anymore important to the management system than other required janitorial work. I'm more than excited about the new housekeeping program we are designing but I wondered if you'd be kind enough to answer a few questions for me?"

I hope that sharing a few reactions, with my responses, may assist those still searching for ways to further increase management's support. Hopefully this will also indicate they now believe - PROFITS REALLY ARE IN ORDER.

— Frank E. Bird, Jr.

Kiyoshi Suzaki received his M.B.A. degree from Stanford and serves on the board of directors of several U.S. companies. He has assisted many Fortune 500 companies in improving their manufacturing operations and in getting their work accomplished better and safer in less time at reduced costs. In his recent popular book, *The New Manufacturing Challenge: Techniques for Continuous Improvement*, Suzaki clearly spells out the step-by-step actions a leader can take to "simplify, combine and eliminate" operations, thus bringing about reduced waste, costs, and improved quality.

The importance this renowned prophet of quality improvement places on housekeeping is shown by the fact that he devotes an entire chapter to the subject early in his book. Chapter 1 is "Eliminating Waste." Chapter 2 is "Back-to-Basics," and deals with "Housekeeping and Workplace Organization." These words by Mr. Suzaki on the first page of this chapter gave early evidence of his deep commitment to this subject:

A west coast company was recently awarded an exclusive right to supply a household product to a large corporation after the buyer visited the factory and found that the people in the factory were committed to housekeeping, workplace organization, quality of work, and efficiency. This was reflected in the morale on the shop floor.

Mr. Suzaki puts an even bigger clinching emphasis in this comment:

"Housekeeping and workplace organization should be among the first steps management takes in improving factory operations."

Here are some additional comments which represent his strong feelings about this subject:

> One of the easiest ways to determine a company's attitude toward improvement activities is simply to walk around the factory and review the housekeeping practices on the shop floor.
>
> The point is that housekeeping ties in closely with many important aspects of management, including operators' morale, management-labor relationships, and level of improvement activities. We should understand the linkage between the level of housekeeping and the amount of defective products produced, the number of machine breakdowns, routing of material flow, inventory level, number of suggestions, level of absenteeism, and so on. And, if we understand the linkage well, each of us should understand better what actions are required to increase improvement activities.
>
> Housekeeping is tied closely to better workplace organization. What we are trying to achieve is not just clean floors and organized racks. The ultimate goal of these activities is to reduce the cost of the product.
>
> Floors and machines should not be cleaned simply for the sake of appearance. Clean surfaces expose problems such as oil leaks and cracks so that corrective action can be taken as early as possible.

While these comments are music to the ears of a believer seeking increased commitment, they reach a beautiful conclusion in the words of Mr. Suzaki's' summary comments.

"Only when these practices are coordinated can a truly first-class operation be achieved."

Some time ago an item, "Japanese Reputation For Productivity Based On More Than Mere Technology," appeared in the business section of the Canadian newspaper *The Globe and Mail.* It was about a speech given at a Toronto seminar sponsored by the Japan Trade Center. Osamu Oishi, a director of the Japanese-owned NTN Bearing Manufacturing Canada, Ltd., of Toronto, outlined his company's efforts to introduce Japanese-style quality and efficiency in a Canadian setting.

> Our management approach was typical of the Japanese method—management by consensus through compromise rather than confrontation through demand. Turning to the task of improving productivity and quality, NTN's management created the concept of 'perfect production' (high productivity with no defects and no accidents) as an objective. To supervise the effort, a joint safety and health management-union committee was appointed. The joint committee was given a mandate to meet frequently and make numerous recommendations for action which could be immediately posted with a management response. The committee was even given the power to stop the production line if substandard measures including safety hazards are observed several times.
>
> For its part, management tried to instill in workers the Japanese penchant for clean and uncluttered work areas. The workers were not enthused, but that was before the President of NTN Bearing Manufacturing got into the act. *He set an example* for the workers *by getting down on his hands and knees and cleaning the floor* himself. The President's action, which included his cleaning

> machinery and removing oil and grease, was somewhat of an unorthodox approach. At first, he was accused of depriving workers of job opportunities. However, several weeks later, workers took the hint and started cleaning their work areas. The result was a remarkable turnaround in the company's fortunes — one that could be repeated in other Canadian manufacturing industries. While at first reluctant, workers soon realized that they were the main beneficiaries of the program.
>
> The psychology used to sell the program was that keeping production machinery and equipment clean and free from oil leaks, dust and dirt meant improved safety, health and comfort for all, particularly the operators. Any sign of oil leakage could be detected immediately, so preventive maintenance was promptly and easily performed. Worker morale improved in the cleaner, more pleasant working environment. In addition, there were other programs centered on innovation and quality. The result was an 80% improvement in quality and a 50% increase in production; industrial accidents were reduced practically to zero. Customer complaints on quality are now virtually non-existent."

Suzaki's and Oishi's comments about housekeeping and order have been supported by observations of Americans and others who have visited operations in Japan (alleged for several years to be the best performers in the world). A few years ago, Robert Hayes, then Professor of Business Administration at the Harvard Business School, researched how various U.S. manufacturing companies were attempting to improve their productivity. In the course of his work, he visited the manufacturing facilities of six Japanese companies. He visited three plants with a group of about twenty-five manufacturing managers from General Electric and visited three other plants on his own.

Professor Hayes reported some of his interesting findings in a *Harvard Business Review* article, "Why Japanese Factories Work." Here are comments related to housekeeping from Professor Hayes' article which amplify the importance order played in what he saw:

> The Japanese have achieved their current level of manufacturing excellence mostly by doing simple things but doing them very well and slowly improving them all the time. (The nail that sticks up is hammered down, says the Japanese proverb.) In the factories that I visited, all the nails appeared to have been hammered down.
>
> In describing some of the ways in which this "hammering" has been done, I shall not discuss the effect of Japanese culture or social norms on management behavior, the distinctive aspects of Japanese management systems or the virtues of Japanese industrial policy. They are all important topics, but all have been the subject of innumerable books and articles. Instead, I will focus simply on how the Japanese manage their manufacturing functions.
>
> The factories I visited were exceptionally quiet and orderly, regardless of the type of industry, the age of the company, its location or whether or not it was a U.S. subsidiary. Clearly this orderliness was not accidental. The meticulousness of the Japanese worker was not, in my opinion, the major reason for the pervasive sense of order that I observed but seemed instead to result from the

> attitudes, practices and systems that plant managers had carefully put into place over a long period.
>
> Sources of litter and grime were carefully controlled, boxes placed to catch metal shavings, plastic tubs and pipes positioned to catch and direct oil away from the workplace, spare parts and raw materials carefully stored in specific areas. The rest areas were centrally located, tastefully decorated and immaculate.
>
> In the factories I saw, the sense of order also resulted from an almost total absence of inventory on the plant floor.... The little inventory I did observe was carefully piled in boxes in specific places around the plant—marked, as were the aisles, with painted stripes.... Along with regular preventive maintenance and constant cleaning and adjustment, machines last longer with reduced rates for use.... I expected to be impressed by the newness of Japanese machine tools compared to those used in the same industries in the United States... but the machines were not really that much newer, they just looked newer and they ran newer.... The Japanese have such trust in the error-free functioning of their equipment that they often load up a machine with work at the end of the last shift and let it run through the night.

The values of cleanliness and order for quality assurance programs are obvious. The reader, like the authors, has probably heard comments from various visiting customers reporting observations, such as, "It's easy to see why your product quality is so good." However, I also remember when our operation was so unsightly that it was embarrassing to have visitors—and I can also remember another customer visiting who connected the dirt and disorder with our high defect rate.

One experience deeply etched in our memory occurred a number of years ago in a heavy industrial plant. A quality control manager was irate about the lack of order in one of the product shipping areas. A tractor trailer backed into the loading zone of this department with eight of its heavy tires rolling over and ruining what was intended to be sold as the world's highest quality polished stainless steel plates. The rework and delays involved with several important items for one of the biggest customers during a slack business period gave birth to an emphasis on orderliness that influenced one of the authors entire management career.

The Japanese 4S Program

Many published articles include comments about the meticulousness of the Japanese and the orderly appearance of their work areas. The previous comments of Professor Hayes fairly well represent and even go a bit further, in his descriptive detail, than most articles. While frequently observed by managers from the Western world, this element in the formula for the great success of quality in Japan has apparently been largely overlooked in its importance to attainment of their improvement goals. Now that tremendous strides have been taken to achieve competitive footing, many more useful factors are coming to light. Of course, one of the biggest sources of knowledge in this regard are the plants in the U.S. managed at the top by Japanese.

The common practice of management idea exchange in executive conferences and the frequency of inter-continental travel have brought mutual insight about important factors in management skills and techniques. A chance reading of materials exchanged with the Japan Safety and

Health Association in 1991 drew attention to a manual of "Successful 4S Movement in Small Japanese Businesses." Since we had been asked several times, to our embarrassment, if we knew anything about the Japanese 4S program, we made it a point to obtain a manual as quickly as possible.

It contained sixteen detailed case histories of industrial locations applying the 4S movement, detailing their key implementation efforts and the results. The following comments on results are taken from one typical case history, Press Working Company C. Please note the references to improvement benefits. The meaning of the 4S's is shown in *Figure 2-1.*

Effects of the 4S Movement

(1) Decrease in Industrial Accidents
The press segment of this company has been maintaining a no-accident record ever since 1974.

This, in turn, resulted in a substantial decrease in the percentage of defective goods, which is one-third less than it used to be. In addition, the operation rate of machinery and equipment had increased, while a significant increase has been registered in the company's sales (about 2.5 times as much as previous levels).

(2) Growing Public Reputation
At this company, a campaign for a cleaner and more beautiful workshop is waged, and all employees call a greeting to one another every morning and evening. Many employees bring flowers raised in their home gardens and arrange them in the workshop. The working environment is now remarkably improved where employees are working in a congenial atmosphere.

As a result, this company is now rated very high among customers and receiving an increasing inflow of orders.

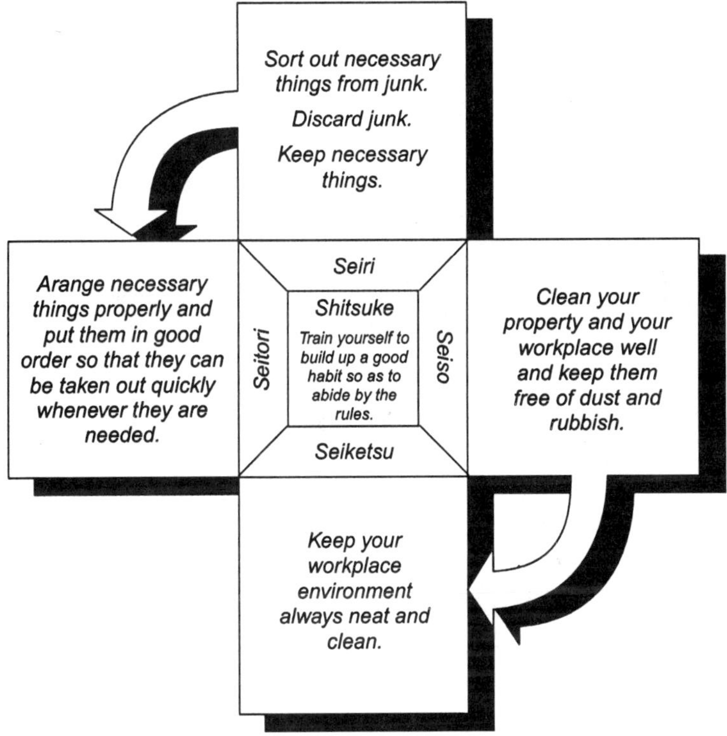

Figure 2-1. The 4 S's

> In job offers, the company is praised by the local public employment agency. Thus the company's public reputation is growing.

After thoroughly reviewing the manual, we were quite amazed at how similar the terminology associated with housekeeping was to our texts published through the years.

Recently we obtained another interesting book entitled *5 Pillars of the Visual Workplace - The Sourcebook for 5S Implementation* by Hiroyuki Hirano from *The Business Reader* publishers. This is an English translation of a Japanese book implementing the 4S or 5S housekeeping program. We have learned the program is described either way. This book is advertised as a unique source book that fully explores the industrial housekeeping system known in Japan as the 4S or 5S program, a cornerstone of any continuous improvement process. The success this program has had in Japan is apparently well-known to Mr. Norman Bodek, the publisher of the English copy. He wrote a Publisher's Message of several pages ahead of the usual book preface. Because of his very strong feeling about the value of housekeeping to quality and productivity, we have printed his message on page 51.

Order Stimulates Quality Prone Attitude

Isn't it also logical and common sense that the performance of anyone in a dirty and disorderly workplace would eventually reflect poor habits and a don't care and sloppy attitude? While anonymous, the question at the start of this chapter states uniquely what the authors have heard dozens of times over the years.

> "The appearance of a place is a self-portrait of the person who works there."

Harry Meyer, a senior executive frequently quoted by the authors puts it another way.

> "Go into any place that is filled with a lot of unnecessary dirt and disorder, and you will never feel like working."

If there is any doubt about the effect of disorder on the average worker's attitude toward work, ask yourself or your spouse the following questions.

- Do you go to a clean and orderly gas station, or to a dirty and disorderly one?

Publishers Message

The book before us is *5 Pillars of the Visual Workplace: The Sourcebook for 5S Implementation.* I consider Hiroyuki Hirano, the author, to be a genius and the 5S's to be the most important step for productivity improvement and safety today. Let me explain what I mean.

Products and processes today involve very close tolerances. Variability must be fully controlled. The physical work environment is critical in the drive for high quality, low cost, and speedy delivery. Think about the Indy 500 in which, thanks to precise teamwork and organization, a car can be serviced in the pits in only 14 seconds. The computer industry has created "clean-room" work environments to produce the required precision and purity in production. Now growing consumer demand for quality products is forcing people from other industries to rethink their workplaces.

Will product improvements see the light of day (or night) in filthy plants? Can we expect people in dismal environments to work at their maximum potential? Can uncluttered minds with fresh ideas function in cluttered workplaces? The answer is obvious — and the solution so simple.

Organization, Orderliness, Cleanliness, Standardized Cleanup, and Discipline are needed. This is what Mr. Hirano calls the "5S's" — simple activities that can be difficult to implement. And this is what is required for companies to survive in the years ahead. Americans too often expect high drama from an idea when what is needed is simplicity — and a solid foundation to sustain good work.

— Frank E. Bird, Jr.

- Do you like to get your car fixed in a clean and orderly garage, or in a dirty and disorderly one?
- Do you go to a clean and orderly barber shop, or to a dirty and disorderly one?
- Do you go to a clean and orderly grocery and meat market, or to a dirty and disorderly one?
- And when you hire people to work for you at your home do you like them clean and orderly or dirty and disorderly?

Doesn't your reaction to these questions generally show how mental attitudes are affected by the appearance of the environment?

Professor Robert Hayes of the Harvard Business School is quoted earlier on his impression of housekeeping in Japan. After one of his trips with a group of industrial leaders, he quotes the response of one of his colleagues that was fairly representative of the group reaction.

> "If you clean up the factory floor, you tend to clean up the thought processes of the people on it too."

Mr. Suzaki, author and industrial leader quoted earlier, points out that poor housekeeping doesn't just reflect on the workers. His statement is emphatic.

> "Housekeeping practices reflect management's general attitude toward work."

Henry Ford, the great inventor and industrial giant, was asked the question, "What would you do if you were called upon to take charge of a business that had failed?" His response was, "No business I know ever went to the wall without first accumulating a vast pile of dirt. The dirt and all that goes with it, untidy thinking and methods, helped to cause that failure. The first thing I would do would be to clean that business up."

Researchers and writers like William James and W. J. Abernathy of Harvard University confirm that industrial revitalization has much to do with greater involvement in human resources development and applied behavioral science. In effect, they are talking about the revitalization and restoration of job pride. Job pride has been described as an attitude, a mainspring that motivates people to do their best, a state of mind that says, "I am important. My job is important. I want to do my job as well as possible." Job pride involves self-respect and self-esteem; respect and esteem for the workplace, the tools and equipment needed to do the work; and respect for one's supervisor, leader and company.

Job pride development is what we do to wind that motivational mainspring to bring out the best in people. It includes all the things we do to influence, encourage and inspire people to peak performance. In that regard, nothing we do, or encourage our people to do, is more important than providing tools, equipment and workspace that creates an image consistent with the mental picture people normally have of an environment conducive to efficiency and excellent performance — a clean, orderly workplace.

Perhaps no analogy about the influence of the work environment on attitude and job pride is more apt than the thought of a dirty, disorderly restaurant. Have you ever walked into a restaurant hungry for a special food and with high expectations for a relaxed, enjoyable meal only to find the entire atmosphere dirty and disorderly? Details are not needed to illustrate the negative effect of dirt and disorder on the digestive process. Likewise, it is just as easy to understand why cleanliness and order in a workplace have such a strong influence on a worker's attitude toward work performance. Numerous management participants in our training programs report increased productivity following an organized program of good housekeeping and order. Improved morale is always reported as one of the biggest benefits, with frequent reports of production increases of 12-16% as a by-product of this attitude change.

Perhaps no example of a clean, orderly plant and its effect on the attitudes of people is better than that of the Alpo Petfoods, Inc., plant in Crete, Nebraska. We believe this plant has one of the finest overall management systems anywhere. Mr. Frank Lothrop, the former General Manager, indicated that prior to the institution of their Alpo safety program (based on the International Safety Rating System), employees' attitudes toward the physical plant environment were substantially different than those they hold today. Mr. Lothrop indicated that at one time the employees had low self-esteem, with as much as a 60% turnover. People were not very proud of their employment. "If anyone asked our employees about their work," he remarked, "they would probably answer, 'Well, I'm working at that dog food plant — do you know of anything better?'"

"Step by step," he continued, "the plant changed radically." It took five years to reach their present level of success. The company experienced a substantial increase in quality and productivity and well over a 70% decrease in plant accidents. Today, this plant has an absentee rate of three quarters of 1% (0.75%) and has been the recipient of prestigious national awards for safety and quality circle innovation. Perhaps one of the most significant indicators of positive change is the fact that 11.2% of the entire employee force is currently enrolled for college credits.

Mr. Lothrop points to implementation of the safety and

housekeeping system as the catalyst that resulted in the present outstanding positive attitude and overall performance accomplishments of his people. Anyone visiting the Alpo plant at Crete, Nebraska, quickly notices the high level of cleanliness and order that characterizes this safety emphasis. Employee interviews confirm that continuous recognition for excellence in cleanliness and order has been a common activity from the start of their outstanding program. The Alpo story supports the fact that a clean, orderly plant experiences improved quality, fewer accidents and creates a workplace free of the environmental hazards to which many experts attribute high absentee rates.

There has never been anything more certain in the minds of the thousands of professionals we have had the privilege to meet than the fact that a clean and orderly shop encourages better work habits. In effect, working in a quality prone environment tends to create a quality prone attitude toward work.

Past Attitudes and Modern Concepts

Understanding the difference between housekeeping as a regular management activity and housekeeping as a janitorial service can well make the difference between opening and closing the door to one of the greatest cost-reducing and productivity-improving activities available to modern business. The benefits of order — in quality assurance, waste identification and reduction, job pride development, productivity, accident control and reduced insurance costs — can only be fully achieved by properly understanding fundamental concepts critical for the development of an effective program.

Historically, housekeeping promotion has been almost exclusively linked with the clean-up associated with janitorial services and a company's safety program. Typical slogans such as "A Clean Plant is a Safe Plant," "Clean Up for Safety," "Clean up; Don't Slip Up," "A Cleaner Place is a Safer Place" and "A Place for Everything and Everything in its Place" have been used for decades and continue to be promoted. Poster catalogs continue to offer the same themes they have been offering for what seems like ages. When one considers the astronomical price of work-related injuries and illness, the focus on housekeeping for its safety values seems entirely justified.

> "A place for everything and everything in its place." It is precisely the literal application of this century-old definition that costs industrial America over 40 billion dollars a year.

However, it does not take much imagination to guess the level of attention that housekeeping and order would be given if the total cost benefits, including those related to safety and health, were considered thoroughly. Past slogans and definitions do not properly reflect the total picture.

Consider for a moment the most widely used and published definition of *order* as related to housekeeping: "A place for everything and everything in its place." It is precisely the literal application of this century-old definition that costs industrial America over 40 billion dollars a year (a sum almost equal to the 1991 national payments of workers' compensation) because of waste caused by disorder.

Finding a place for such items as the extra bolts or brass fittings found after a pipe fitter's repair, the half-filled container of solvent, those three six-foot 2 x 4's left last week by the carpenters or the two bags of cement that were never picked up by the masons—in other words, finding a place for everything and then putting everything in its place—could be a drain causing continual big losses for any organization.

It staggers the imagination when one adds the costs of major items such as machinery, equipment and major quantities of supplies that are totally unnecessary for the areas where they are located. This costly disorder occurs as the habits of individuals spread to broad general practices and eventually reflect a lack of an organizational control over storage. Making an "attic" of any plant location has long been and continues to be an accepted housekeeping practice because any location can be made to appear very clean and orderly this way. By creating an attic, we have made "a place for everything and put everything in its place." The fact that many costly items may be totally unnecessary, unneeded and wasteful is not the point. The objective is to make things appear orderly, whether or not they really are:

Housekeeping and order as important management activities and catalysts for improved quality and productivity is best realized by a new definition of order that reflects this philosophy. One way to accomplish this would be to build upon the most widely used definition of *order* (a place for everything, and everything in its place) and benefit from the lessons of the past. By simply adding the word *proper* in two positions of the old definition, we have a new, more meaningful principle: "A proper place for everything and everything in its proper place." This modification indicates clearly that only those items proper and/or necessary to the location should be placed there. However, correcting long-established patterns of behavior may not be as easy as making these simple changes.

Since the purposes of this chapter are to revitalize interest in the benefits of good housekeeping and order *and* to present a methodology for achieving success, we would like to present another definition of order that has long been an important part of our own management philosophy.

Some years ago at a safety conference in Baltimore, Maryland, we heard one of the most impressive speeches of our business careers. The keynote speaker was Harry Meyers, a senior executive with a major company, and his subject was "The First Law of Good Work - Be Clean and Orderly." We vividly recall that this man's confidence in and enthusiasm for his subject commanded respect from the very start. He captured everyone's interest at the start by announcing that he was going to address a subject that could substantially alter the profitability of any company represented in this large audience. We found ourselves making detailed notes and writing as many direct quotes and comments as possible since Mr. Meyers had already convinced us that he knew what he was talking about and would say something of great value. In short order, he walked away from the podium and requested that participants offer definitions or slogans related to housekeeping or order. As participants offered definitions like "A clean plant is a safe plant," his responses were, "That's probably true" and "Yes, I believe that one, too." When someone finally said, "A place for everything and everything in its place," his voice and body language captured special attention as he stated loudly and clearly, "That's the one I've been waiting for."

He repeated, "A place for everything and everything in its place," followed by the blunt comment, "Don't you believe it! Any organization that believes and practices that one will have everyone creating attics throughout the operation; your inventory costs will soar. Safety and quality will be hurt badly, and the only thing that will suffer more than these three will be the profitability of your company."

He then cited example after example of totally unnecessary items either left by service employees or left over from some previous use or application. He continued by explaining that by making a place for each item and neatly storing it, the general appearance of an area could be perfectly clean, even orderly, to someone unfamiliar with the area. But to a manager who first knows his or her operation and second knows the true meaning of order and whether or not each item is necessary where it is located, these practices constitute costly disorder.

"Let me give you a definition of order," Harry Meyers stated clearly and deliberately, "that if applied religiously can become the foundation for total business efficiency in your management career."

> "A place is in order when there are no unnecessary things about, and when all necessary things are in their proper places. "No" in this sentence means none! Not any! Not even one!"

Mr. Meyers described how *order* in this new context would have a dramatic effect on a long list of organizational activities. His comments were buttressed by example after example of the proven results of a program of order.

Either of these two valuable definitions of order could be used in any housekeeping program with effective results. You may have guessed, from the detailed explanation, the authors favor the definition of Mr. Meyers.

Practical Implementation Steps

OK, so logic, common sense, expert testimony and successful practice clearly point out the fact that a clean, orderly workplace creates a quality prone environment. Isn't this truth so widely known and accepted that it's no big deal? Isn't everyone trying to keep their shops clean and orderly? Is there something new that management is generally unaware of?

You bet there is! Managers have been searching for years for a method to objectively rate housekeeping so that all locations in a plant are treated fairly. This would take the subjectivity out of methods used so that a group could have some confidence in its own efforts as well as comparing its efforts to those of others. The search for such a rating method has been as elusive as the search for information on the Fountain of Youth.

Not knowing that there is an objective, statistically valid way to rate housekeeping has kept order from lending its maximum values to our improvement programs. For a worker or leader to maximize benefits from a quality prone environment, they have to first clearly know, without doubt, what the standards of order are, and that they are going to apply consistently as everyone strives to gradually achieve 100% effectiveness. Nothing is more demoralizing to a worker or a leader than to put forth dedicated efforts in any program only to be rated or evaluated less than his/her efforts deserve. This is exactly what has happened time and time again with housekeeping rating systems of the past.

Jorma Saari, an internationally respected safety researcher and Director of the Institute of Safety and Health in Finland, recently conducted a detailed study of housekeeping with a team of his colleagues. His findings and conclusions were more than interesting. When well-known, they will have a dramatic effect on those organizations who accept that when order is managed it can make a very positive effect on any organization's continuous improvement program. Dr. Saari's findings were promptly incorporated into over 100 various plants in Finland and continue to be studied by his team. One of the authors of this text felt the findings of his study significant enough to quality, productivity and safety to travel to Finland to visit plants using the program and talk directly with members of the team who participated in the original research.

Here are just a few of the most interesting findings of this first in-depth scientific study of housekeeping taken from Dr. Saari's chapter on "Scientific Housekeeping Studies" in the book *Profits Are In Order:*

A. Improving housekeeping is profitable, even in a company with a fairly high initial housekeeping rating.

B. Feedback to workers on the floor following a rating is just as motivational to stimulate continued interest as prizes or rewards for attainment.

C. Ratings should be based on a few standards selected by a representative team of workers from each area to be rated.

D. Housekeeping ratings should be based on purely objective decisions of compliance or noncompliance with standards.

E. A baseline rating should first be established using the standards established. Future ratings are then compared to the baseline.

F. Self-comparison ratings are most effective in producing results.

G. Over time, ratings can be less and less frequent to maintain satisfactory levels of housekeeping.

The following suggestions, combine the very latest concepts on the use of regular measurements of housekeeping conditions with Dr. Saari's major thoughts and findings.

1. **Program Coordinator**
   The creation of a program is best accomplished when all necessary work is coordinated by one facilitator, ensuring the homogeneous attainment of objectives throughout the entire location. Historically, such an assignment would probably be directly associated with

the safety/health function. While there is no major reason for this association to end, we hope that readers of this are now convinced of the many benefits of order. While housekeeping alone has some effect upon accident control, the greatest benefits of order are probably its overall benefits to the management system. As Dr. Saari writes, "The reduction in accidents we found is much more a result of better industrial relations than of order, per se. This is not to demean the benefits to safety, but to emphasize that safety is only one of the benefactors of good workplace order." Project coordination by a quality assurance or production leader would be just as appropriate as coordination from the safety function.

2. **Area Team Appointment**

Depending on size, it may be appropriate to break the site into smaller units for rating purposes. Since housekeeping and order should be managed as any other management activity, the normal operational division of the site would probably serve this purpose well. A larger location might have divisions under senior executives, with departments or subdivisions under managers or department heads. A housekeeping system is frequently established at the department or subdivision level.

Experience teaches that areas with employee groups from 10-30 work quite well for a program of order; it may be wise to break large departments into small groups. Departmental and division ratings could be established by using a mean score of the component areas/groups. Since participation is a vital key to success in any management system, ownership through involvement is best achieved by using a team from each area to establish the specific standards for measurement of that area. While the size of the team could vary, experience teaches that teams of 3-5 members who are quite knowledgeable about activities in the specific area work best.

3. **Setting Key Area Goals**
   One of the first functions of the team should be to establish key housekeeping goals for their area. Key items should be selected to represent the 10-15 most desired housekeeping goals for the specific area. The key items could relate to a variety of goals that, when attained, would have the greatest impact on safety, quality and production. The specific standards for measurement of the area will be based on these goals/targets. Depending on the nature of the questions, the goals could be similar to those in *Figure 2-2.*

4. **Establishing Area Performance Standards**
   Establishing area performance standards is unquestionably the most important function of the team since it creates the basis for the observations that will objectively measure and evaluate the level of housekeeping and order for their area.

   While the number of standards for an area may be as few as 1-15, they should be applied to each work position, storage and other specific locations within the area to fix responsibly and identify exact location of any noncompliance. To facilitate ease of scoring and objectivity, there should be at least 100 or more items to rate in each area. The rater only stops whenever an item of noncompliance appears on a safety walk through. Rating becomes easier as the evaluator becomes familiar with the inspection route of each area.

   - Convert each goal agreed upon to a standard.
   - Standards should be written as clear, positive

Typical Housekeeping Goals

MACHINERY AND EQUIPMENT

a. Must be clean and free of unnecessary material or hangings
b. Must be free of unnecessary dripping of oil or grease
c. Must have proper guards in place and in good condition

STOCK AND MATERIAL

a. Must be properly piled and arranged
b. Must be loaded safely and orderly in pans, cars and trucks

TOOLS

a. Must be properly stored
b. Must be free of oil and grease when stored
c. Must be in safe working condition

AISLES

a. Must be provided to work positions, fire extinguishers, fire blankets and stretcher cases
b. Must be safe and free of obstructions
c. Must be clearly marked

FLOORS

a. Must have surfaces safe and suitable to work
b. Must be clean, dry and free of refuse, unnecessary material, oil and grease
c. Must have an adequate number of receptacles provided for refuse

BUILDINGS

a. Must have walls and windows that are reasonably clean for operations in that area and free of unnecessary hangings
b. Must have lighting systems that are maintained in a clean and efficient manner
c. Must have stairs that are clean, free of materials, well lit, provided with adequate hand rails and treads in good condition
d. Must have platforms that are clean, free of unnecessary materials and well lit

GROUNDS

a. Must be in good order, free of refuse and unnecessary materials

Figure 2-2

statements, whenever possible, not more than one sentence long. Make them as crystal clear and understandable as possible.

- Standards should clearly define what is acceptable.
- Standards should be clear and understandable to observers from other areas.
- Standards should be realistic and attainable.

Here are some typical standards:

1. **Storage of Product and Material**
   Product and material must be stored in an orderly manner, with only needed material in the shop,with only parallel and perpendicular storage and locations clearly identified.

2. **Condition and Storage of Tools, Equipment and Vehicles**
   All tools, equipment and vehicles must be clean, in proper condition and stored in designated areas except when actually in use.

3. **Disposal of Shop Waste**
   All shop waste must be discarded directly into adequate containers, which must be identified and not more than three quarters full at any time.

   Once standards for rating housekeeping conditions in an area are agreed upon, the team should also recommend any necessary provision of specific hooks, racks, cabinets or provisions for proper storage of tools and appliances. This is one of the reasons for appointing a program coordinator, to ensure as much consistency as practical from area to area and to facilitate such activities as purchasing, maintenance and overall program continuity.

4. **Eliminating Specific Barriers to Goal Attainment**

The team members of each area will know the specific barriers that ought to be removed/corrected before the rating system is put into operation. These barriers could require such things as providing proper refuse containers and arranging for their continuous care. Other actions might be providing for necessary hooks, racks, cabinets or other storage provisions for tools and appliances. Instructional signs, labeling of emergency equipment and painting of lines or areas designating keep clear limits are additional actions often needed.

5. **Establishing Baseline Ratings**
This is one of the most important facets of the program. It will take time but success of the entire program depends on the accuracy with which baseline ratings are determined. To have a fairly accurate picture of the baseline to which all future ratings will be compared, at least 3 ratings should be made using the new standards and rating form for each area. Ratings to establish the baseline should be unannounced and conducted independent of any announced inspection or visit. It is best if these ratings can be done over 2 or 3 months to get a true baseline. Naturally an average rating would constitute the baseline. Whatever the number determined, the same number of ratings should be made in each area. The baseline rating for each area should be established before any program explanation such as that described in item 7 is communicated. All ratings should be unannounced. It is best to do ratings on all shifts. The attached form (*Figure 2-3)* will serve to illustrate features you may want to incorporate in your rating form design.

6. **Communicating Program to Everyone**
After the baseline rating for each area has been established, one or more of the regular safety meetings can

Housekeeping Rating Form

April, 1995
Date

Robert Stevens
Evaluator

ABC Department
Area

49.6
Present Housekeeping Rating

Item Rated	Standards/Rules	C	N/C	Comments
1. Drill press #26	Must be free of unnecessary material or hangings; oil or grease	✓		
2. Drill press #26	Must have proper guards in good condition	✓		
3. Drill press #26	Operator's tools must be in good condition, properly stored		✓	3 tools with unsafe handles or edges
4. Drill press #26	Stock and material excess around press must be properly piled and arranged		✓	Pile of blanks scattered, not in box
5. Drill press #26	Floor space around press safe and free of unnecessary material, oil and grease	✓		
6. Stock and material Area 1	Piling and/or loading arrangement safe and orderly			
7. Stock and material Area 1	Aisleways free-of-obstructions and clearly marked		✓	Stock in aisles (3 locations)
8. Stock and material Area 1	Fire exiting fire blanket and stretcher provided, free of obstructions, in proper condition	✓		
9. Stock and material Area 1	Floor space safe, free of unnecessary material, oil and grease	✓		
176. Work station #6	Tools in safe condition properly stored	✓		
177. Work station #6	Stock and material properly piled and arranged		✓	3 unsafe piles (height and method)
178. Work station #6	Floor space safe, free of unnecessary material, oil and grease.	✓		
179. Work Station #6	Fire exiting, fire blanket, stretcher provided, free of obstructions, clearly marked, safe		✓	Fire blanket moldy and unclean
180. Main aisleway	Free of obstructions, clearly marked, safe	✓		
181. Grounds around building	Good order, free of refuse and unnecessary material	✓		
	Total Points for compliance (C)	85		
	Total Points for noncompliance (N/C)		96	
Present Housekeeping Rating = (total possible points (181) divided by total points for compliance (C) multiplied by 100.)		46.9%		

Figure 2-3

be utilized to communicate standards and program aspects to all employees. Program success is best achieved if a team member is motivated and prepared to make the final presentation to fellow employees. Part of the presentation should be rekindling their ownership in the program. Further of course, ratings will be made on the standards they created or selected. The team member or members making the presentation should be challenged to motivate fellow workers to cooperate in making their area the best. Representation by upper management at these training sessions certainly demonstrates leadership interest.

Regular Ratings and Their Evaluation

Peter Drucker has said, "The measurement used determines what one pays attention to." These comments on measurement and evaluation are as valid today as they were centuries ago when Galileo stated, "Count what is countable, measure what is measurable and what you can't measure, make measurable." The objective of a modern housekeeping rating system, then should be to provide leaders with a number that represents the percentage of total compliance with established standards for the area evaluated.

Dr. Saari, mentioned earlier, discovered in his research the critical importance of the feedback system for program success. There are several methods that have been used by other leaders.

Figure 2-4 presents a feedback form exclusively devoted to housekeeping ratings/evaluations that would be best suited for feedback to management. Feedback to people on the floor is extremely important for program success and a special effort should be made to design the most effective graphic approach to accomplish this.

Quarter

Year

Comparative Housekeeping Ratings

Area Rated	J	F	M	A	M	J	J	A	S	O	N	D	BR	TYD	LY	TQ	PPY
Averages																	

RATING LEGEND

BR - Base Rating
LY - Last year
PPY - Performance position year to date
TYD - This year to date
TQ - This quarter

Figure 2-4

While *Figure 2-5* offers an easy way to display results for people on the floor by simply using numbers, bar graphs and trend charts have a great psychological appeal by presenting the complete comparative rating history. Whatever feedback display of housekeeping ratings is used, it should be large enough to attract attention from a reasonable distance on the floor. Simplicity of design and the use of attractive colors help achieve this objective.

Earlier in this chapter, we mentioned that one of the authors had visited the Institute of Health in Finland to talk with people who knew the practices of well over 100 locations using the objective housekeeping system developed by Dr. Saari. One of the most important facts he learned was that while 50% were simply using feedback displayed in the workplace as the only form of recognition, surprisingly enough, there was no discernible difference among the results of the two groups. This fact emphasizes the need to plan the use of a feedback system well.

Department Housekeeping Ratings

Base Rating	Present Rating	Best Rating	Year to Date
45	68	74	71

Figure 2-5

Achieving excellence in any management activity is a function of time (never-ending improvement) and the level of leadership that prevails. The effective leader constructively corrects deficiencies in the system as the measurement and evaluation procedures document noncompliance with housekeeping standards. Hopefully, this chapter has provided enough knowledge and motivation for the program coordinator to identify both performance deficiencies and strengths and to see them both as growth opportunities to achieve optimal desired results. In turn, we feel certain your program success will unquestionably prove that good housekeeping can be a key to developing a quality prone environment.

Satellite Construction Services

John A. Browning
Executive Vice President

Jack Oswald
Division A Manager

Dear Jack:

We will shortly be kicking off a program that will be referred to as "Operation Order." An important part of this will be the introduction of a new system to objectively measure the effectiveness of housekeeping and order of all operating areas of our plant on a monthly basis. The measurement system comes from Europe where it was scientifically tested in well over 100 plants.

The exciting thing to me about our new emphasis on "order" is that we have always connected good housekeeping with safety and health but now realize it will have an enormously important effect upon our quality assurance and productivity efforts. In effect, I view this program as a double-edged sword of opportunity to demonstrate increased concern for our people while at the same time improving our profit performance. I'm sure you will agree that our total dedication to this effort is sound business planning. You will shortly be receiving some pre-session reading material from Paul Davison, our "Operation Order program coordinator. Please read it carefully before the session date to be announced in Paul's letter.

I look forward to hearing your reactions and suggestions at the kick off training session. I know you will contribute to the program's complete success. Thank you in advance for your cooperation.

Sincerely,

John A. Browning
Executive Vice President
Operations

Typical program introduction letter

Selected References

Bird, Frank E., Jr. *Profits Are in Order.*, GA: Institute Publishing, 1992.

Bird, Frank E., Jr. and George L. Germain. *Practical Loss Control Leadership*. GA: Institute Publishing, 1992.

Casebook of Successful 4S Movement in Small Japanese Business. Tokyo, Japan: Japan Industrial Safety and Health Association, 1990.

Hayes, Robert H. "Why Japanese Factories Work." *Harvard Business Review.* July-August, 1981.

Hirano, Hiroyuki. *5 Pillars of the Visual Workplace.* Portland, OR: Productivity Press, 1995.

Saari, Jorma. "Scientific Housekeeping Studies." *Profits Are in Order*. Atlanta, GA: Institute Publishing, 1992.

Suzaki, Kiyoshi. "The New Manufacturing Challenge." *Techniques for Continuous Improvement*. NY: The Free Press, 1987.

Chapter 3

Safety Applications of Economic Axioms

"Management can only justify its existence and its authority by the economic results it produces."

1954 — Peter F. Drucker

"Until a business returns a profit that is greater than its cost of capital, it does not create wealth; it destroys it."

1995 — Peter F. Drucker

INTRODUCTION

Applying economic as well as humane aspects of accident control has been extolled as one of the best management motivators by many safety leaders, from great pioneers like

Heinrich and Blake to contemporaries like Petersen, Pope and Grimaldi. It remains the number one method to win the executive support so universally sought by safety leaders. In spite of its long advocacy, leaders everywhere seem to agree that accident control remains one of the few avenues in many organizations for substantial cost improvement.

An article entitled, "Loss Control Creates Value and Competitive Advantage," by Dr. Douglas Clark and Per Olaf Brett in the *DNV Forum* (1995, No. 1) by Det Norske Veritas, starts with these interesting comments: "The enormous impact which uncontrolled losses can have on an organization can be seen from a study done at Old Ben Mine in Australia. When the identified costs of rework were added to the cost of injuries, illness, waste and property damage, the amount was equal to 25% of total operating costs."

"Our conversations with a number of maintenance managers indicate that on the average, 40% of a company's maintenance costs are due to abuse, misuse and accidental damage.

"A major beverage company in the USA discovered recently that its accident losses exceeded all the profits from one of its largest divisions. When a national electric utility company was prodded by a DNV consultant to identify its costs for accidental property damage and waste, the figures were so large they were promptly hidden for fear of creating a national scandal. Loss of this magnitude must be taken into account in considering competitiveness. Some organizations have discovered that a proper approach to loss control can make a significant contribution to the bottom line and thus improve their value creation and competitive position. One example is Syncrude Canada, a tar sands operation in northern Alberta, which meets about 12% of Canada's total petroleum needs. Ralph Shepherd, who was chief executive officer at the time, stated in a 1989 speech that management had made a firm decision to place safety and reliability at the top of their priorities. He defined reliability as 'consistency of performance,' being able to trust someone or something to perform when you need it. He defined safety as 'the condition of being free from hurt, injury or loss Shepherd continued, 'While reliability represents the more mechanistic, functional part of the equation, safety embodies the humanistic, moral side. At Syncrude we have brought the concepts together to reflect our belief in the fundamental relationship between them. As a result, lower maintenance costs and increased production helped Syncrude significantly knock down the cost of producing synthetic crude.'

"One reason was that, 'When people believe in and are committed to working safely, they extend the same meticulous approach to the way they maintain and operate equipment as they do to the way they safeguard their well-being. At Syncrude one action is merely an extension of the other.'"

Another relevant reference to accident costs appeared in the April 1995 issue of *Occupational Hazards* entitled "Investing in Safety" by Ivan Zenker CSP, CIH. It began with these rather shocking and enlightening words:

"In 1993, the cost of work-related accidents in the U.S. totaled $111.9 billion according to the National Safety Council estimates. That staggering sum is equal to a quarter of every dollar of 1993 pre-tax corporate profits. Thus, $111.9 billion loss represents the penalty that American industry is paying for a seriously flawed risk-management strategy."

Assuming the reliability of Mr. Zenker's calculation, the

25% of our national pre-tax corporate profit relates to those accidents involved with injury but, does not include the gigantic injury related liability that has caused some companies to declare bankruptcy. Furthermore, it staggers the imagination to guess what the lost profit would be if the more frequent and costly no-injury accidents discussed earlier in the British APAU studies were costed in.

With these examples to substantiate the effect of accidents on an organizations bottom line, why do the majority of top managers of companies around the world not recognize this fact so critical to cost improvement and competitive ability in today's markets? The answer to this question is well stated in the Clark and Brett article referred to previously.

1. They do not appreciate the magnitude of the costs and waste involved.
2. They do not believe there is any practical means of controlling the costs, and
3. They perceive loss control activities as expense items necessary only to meet government regulations and protect employees.

The truth of this answer is underlined in a comment given by the CEO of a major U.S. corporation at a National Safety Conference, which expressed the perception he knew existed with many of his CEO colleagues. He lays the opinion of many management leaders squarely on the line with these words:

"I think it is important for you to know what the typical CEO thinks about items that are important to him and how he feels about items that are important to the organization Safety is not one of these items, it ranks very low on the totem pole. It ranks low because the CEO is interested in cost; he is interested in productivity and he is interested in return in investment."

This attitude reflects the probable reason that management commitment is such a widely sought after commodity by so many safety leaders. The authors know that attitudes are usually not changed overnight. For this basic reason, the remainder of this chapter discusses economic axioms that, if persistently applied over time, can be major factors in bringing about a realistic perception of accident costs and the substantial improvement to be realized when adequate executive support is given to the safety activity.

1. **The Management Control Axiom:**

   The control of accidents is the best way to control their cost and management has the bulk of this control.

Safety Applications of Economic Axioms

1. The management control axiom.
2. The axiom of economic association.
3. The axiom of the critical few.
4. The key advocate axiom.
5. The past forecasts the future axiom.
6. The mutual interest axiom.
7. The economic priority axiom.
8. The vested interest axiom.
9. The axiom of adequate evidence.
10. The axiom of dimensional value.

It's rather amazing the number of executives who continue to believe that employees and their unsafe acts cause most accidents. In reality those who keep abreast of modern accident control concepts know that 80-90% of accidents are the result of deficiencies in the management system. Only 10-20% can be attributed to pure employee error. Anyone familiar with Deming's or Juran's error removal philosophy knows this same relationship of management system to employee mistakes holds true not only in quality control but in all activities of the management system. Getting managers to accept the premise that they can control accidents is equally as important as the fact that accidents are costly events.

One major reason is that the very definition of accident used in most safety training programs does not imply the reality of control. On the contrary, most widely used definitions contain words like unplanned and unexpected, which reinforces the common perception that these events are not controllable.

Some time ago the "Locomotive", a respected journal of Hartford Steam Boiler Insurance Company, contained a short article entitled "Are Accidents Unexpected?" This one-page editorial pointed out that the Hartford had reviewed 11,000 plus accident investigations they conducted in one year and came to a conclusion that the use of the common term unexpected in accident definitions should be questioned. They cited the common occurrence of lack of maintenance, no procedure for periodic testing of controls, absence of education or training program for operators, no emergency plan or shut down procedure, improper installations, no use of maintenance records or use of logs in plants large and small. The editorial ended with the question, "How can anyone refer to these happenings as being unexpected?"

Accident investigations, HSE training, and all other normal contacts with leaders at all levels should continually strengthen and reflect this key axiom: The control of accidents is the best way to control their cost and management has the bulk of this control.

2. **The Axiom of Economic Association:**

   A manager will usually pay more attention to information when expressed or associated with cost terminology.

There is probably no quotation that better highlights management's keen interest in costs than the often used statement of Peter F. Drucker on the first page of this chapter. The number of similar statements are legend, such as a comment made by I. W. Able, former President of the United Steelworkers of America: "The basic need of every company is to make a profit. Only then can it provide jobs and earnings for employees."

While this seems to be a universal truth, it's rather surprising that safety practitioners do not capitalize on its educa-

tional and motivational power more than they do. Perhaps failure to utilize costs more frequently and effectively for drawing attention to actual and potential accident costs is more not knowing how to go about it than failure to accept the general tenet that "money motivates management."

There are three important caveats experienced safety leaders would point out: (1) Use costs properly and often enough to make the point that control of accident loss can make a significant contribution to profit, but not so often that people become deaf to its message. (2) Be very conscious of the fact that you don't want to impress anyone that the organization is more interested in costs than people. (3) The best use of cost data available to everyone is in damage or non-people applications.

Here are some occasions when costs could be appropriately communicated:

1. Whenever a property damage accident occurs with costs that are excessive, get good estimates of the damage repair and replacement costs. Use this in your regular safety communications. A Polaroid picture, to show the costly results, will also reproduce quite well on most modern reproduction machines. Use property

damage widely in accident costing whether or not you have a full property damage program in operation. It could very well be the start of one.

2. Near loss incidents possess great opportunities to discuss potential loss.

3. Housekeeping that has reached the state of disorder frequently creates unnecessary costs to discuss. Photography can add to the strength of this communication as well.

4. Detection of hazards that could cause injury/damage presents great opportunities to discuss potential costs. You may find it helpful to use several examples in this type of discussion.

5. A significant number of safety leaders have used the average costs of various types of injuries in a variety of ways. One of the most popular ways is to cost all injuries by department and/or division periodically. An average cost is estimated for each type of injury and used whenever actual costs are not readily available. Interest will lag if these costs are not used on a regular,

timely basis. This is the role that the estimated average cost plays until actual costs are known and available.

The average cost of each accidental injury type can be estimated by using this formula and the data that follows:

A x B + C = estimated average cost per injury type.

A = Estimated time lost per injury in hours or days.

Estimate the time involved with each type of injury for travel to medical facility, time of people involved in investigation, delay of others, related administrative activities, etc. Remember you establish this time with a team of representative (supervisors/leaders) personnel. It is helpful to make some actual time estimates for injuries to advise the team. The times given in *Figure 3-1* are the authors' estimates, based on their knowledge and experience.

Estimated Time Losses		
Job Related	Hours	Days
Strictly minor	2	0.25
Medical	5½	0.68
Vehicle	4	0.50
Disabling	16	2.00+
Death	*160*	*20.00+*

Figure 3-1

B = Hourly $ rate for employees (National average costs per industry are shown in *Figure 3-2.*)

Average Hourly Earnings	
Source of 1994 information, U.S. Bureau of Labor Statistics, Bulletin 2445 and Employment and Earnings.	
Private Industry Group	Average Hourly Earnings, 1994
Mining	11.12
Construction	14.89
Manufacturing	14.69
Transportation	13.88
Wholesale trade	12.01
Retail trade	7.49
Finance, Insurance	11.83
Services	11.07

Figure 3-2

C = National average cost for workers compensation paid for type of injury (see *Figure 3-3*). Simple first aid could be an exception, to be estimated locally with consultation of doctor or other responsible person. Estimated 1993 workers compensation benefit payments are contained in the 1995 issue of *Workers' Compensation Insurance: Profiles of the State Systems.* This publication can be obtained from the Alliance of American Insurers, 1501 Woodfield Road, Suite 400 West, Schaumburg, Ill. 60173-4980. The cost for strictly minor first aid cases is an estimation by the authors, but not necessarily intended to be used as your estimated cost.

Median Cost Per Case by Type of Injury, 1993

1995 Alliance of American Insurers

Fatalities	Permanent Total	Permanent Partial	Temporary Total	Med. Only	Strictly Minor First Aid
$114,331	$178,907	$25,049	$1,914	$324	$150

Figure 3-3

6. Use average costs or actual costs as available from insurance or risk management personnel and transfer these costs to the budgets of various departments. If official charge back to budgets does not seem practical, you may still want to use average costs to keep managers fully aware of these controllable and unnecessary accident expenses occurring in their area. The program could also include an annual budget reduction objective for each department, whether or not you are actually charging costs back to department budgets. Of course, the reduction of injuries and their costs will largely be determined by increased safety improvement activities. You could use a cost average for each type of injury, developed locally and largely based on national averages presented in section 5 above, in addition to your estimate of first aid only cases.

7. Keep management well-informed on the costs involved with compensation trends and such items as fines and penalties for lack of compliance. In addition to using this information in your own printed communications, circulate articles from journals or newsprint that share outstanding accident cost data.

   The National Average Incurred Costs for Part of Body Injured, Nature of Injury and Cause of Injury are listed in *Figure 3-4.* Use these in various promotional programs to motivate interest in such activities as the introduction of new protective equipment projects or to justify the related promotion expenditure required. This data was summarized and can be obtained from *Workers' Compensation Claim Characteristics,* available from the National Council on Compensation Insurance, 750 Park of Commerce Drive, Boca Raton, FL. 33487.

NATIONAL AVERAGE INCURRED COSTS Detailed Claim Information							
By Part of Body							
Ankle	Arm	Chest/ Internal Organs	Face	Foot/ Toes	Hand/ Fingers	Head/ CNS	Hip/ Thigh/Pe lvis
$6,124	$11,218	$7,439	$6,346	$5,837	$6,294	$23,299	$10,664

By Nature of Injury				
Amputation	Burn	Carpal Tunnel Syndrome	Contusion/ Concussion	Fracture/ Crushing/ Dislocation
$14,218	$8,410	$12,731	$9,570	$13,983
By Cause of Injury				
Burn	Caught in Between	Cumulative Injuries	Cut/Puncture/ Scrape	Fall/Slip
$8,202	$9,798	$11,816	$5,486	$12,838

Figure 3-4

8. One interesting and common use of accident costs is to translate them into the dollars of sales at various profit margins necessary to make up for them. *Use Figure 3-5* as your guide. Last year's profit margin for your company or industry, can be applied several times during the year without current profit determination each time it is used. See *Figure 3-6* for examples of profit margins for various industries.

In Times of Keen Competition and Low Profit Margins, Loss Control May Contribute More to Profits Than an Organization's Best Salesmen

It is necessary for the salesman of a business to sell an additional $1,667,000 in products to pay the costs of $50,000 in losses from injury, illness, damage or theft, assuming an average profit on sales of 3%. The amount of sales required to pay for losses will vary with the profit margin.

ACCIDENT COSTS	PROFIT MARGIN				
	1%	2%	3%	4%	5%
$ 1,000	100,000	50,000	33,000	25,000	20,000
5,000	500,000	250,000	167,000	}125,000	100,000
10,000	1,000,000	500,000	333,000	250,000	200,000
25,000	2,500,000	1,250,000	833,000	625,000	500,000
50,000	5,000,000	2,500,000	1,667,000	1,250,000	1,000,000
100,000	10,000,000	5,000,000	3,333,000	2,500,000	2,000,000
150,000	15,000,000	7,500,000	5,000,000	3,750,000	3,000,000
200,000	20,000,000	10,000,000	6,666,000	5,000,000	4,000,000

SALES REQUIRED TO COVER LOSSES

This table shows the dollars of sales required to pay for different amounts of costs for accident losses, i.e., if an organization's profit margin is 5%, it would have to make sales of $500,000 to pay for $25,000 worth of losses; with a 1% margin, $10,000,000 of sales would be necessary to pay for $100,000 of the costs involved with accidents.

Example of using specific products sales necessary to make up for the cost of a disabling injury ($1,914) in a department store operating at a profit margin of 2.8%: Our SOS Department store must sell 33,000 pair of quality men's hose to make up for a bad fall from a stepladder by salesperson #2208 last week.

Figure 3-5

CORPORATE SCOREBOARD Data obtained from Business Week/March 6, 1995	
Industry Segment	Composite 4th Quarter 1994 Margins %
Aerospace and Defense	2.8
Automotive	4.4
(a) Tire and Rubber Group	5.4
Chemicals	6.5
Consumer Products (Apparel)	4.7
Discount and Fashion Retailing	2.8
Electrical and Electronics	6.9
Food	2.1
Fuel	4.4
Housing and Real Estate	4.7
Manufacturing	6.2
Metals and Mining	6.7

Figure 3-6

3. The Axiom of the Critical Few:

The majority (80%) of any group of effects is produced by a relatively small (20%) number of causes.

This axiom is based on a widely accepted statistical premise that in the field of economics approximately 20-25% of the items account for 75-80% of the total value, e.g., 25% of the people in the world own 75% of the world's wealth. This concept is based on the observations of an Italian economist Vilfredo Pareto, who set the axiom down as an economic Law in 1906. In his honor, it became widely known as "Pareto's Law."

Management authorities through the years have repeatedly substantiated and broadened "Pareto's Law." Perhaps no one emphasized this more than the well-known quality improvement consultant and Fellow of the American Management Quality Improvement Association, Joseph M. Juran, who in 1954 describe this axiom in his lectures and writings as "universal for management planning and control." Other terms such as "vital few" and "trivial many," "main issue" and "critical few" have been used in a wide variety of management applications.

A wide range of computer software is available to store and analyze accident data, including costs. Identification of the critical few, assists the safety practitioner, risk manager, or insurance company in focusing on those causes and costs that provide the greatest opportunity to cut the cost of risk with the least expenditure of time and energy. These become the immediate targets for concentrated management effort.

A very dramatic illustration of this principle can be found in the last national average for types of injuries (1991) listed in the 1995 copy of *Workers' Compensation Insurance: Profile of the State Systems* from the Alliance of American Insurers. Average state compensation costs by injury type are shown in *Figure 3-7*. Of course, this axiom with its broad application and potential benefits to leaders can also be aimed at losses from theft, fire or any other loss area.

Extensive use of this axiom is made in property damage and waste control as described in our book, *Practical Loss Control Leadership*. An inventory is made of all damaged items repaired or replaced annually in shops or key field repair areas. An average cost of repair and/or replacement costs is obtained from the area supervisor/leader. With these figures, it is not difficult to estimate quite accurately the critical few items based on costs. By assigning these items to project teams the cause of the problems can be determined and corrective measures taken. A frequently used form, to obtain data from repair centers, is shown in *Figure 3-8. Figure 3-9* shows the items from an actual study. Application of the Axiom of the Critical Few highlights the items which will most justify attention.

National Averages for Types of Accidents		
Injury	% of Total Injuries	% of Total Costs
1. Medical Only	82	14
2. Temporary Only	13	13
3. Permanent Partial	5	66
4. Permanent Total	2/5 of 1	4
5. Fatality	2/5 of 1	3

Note: Since figures have been founded off the % of total injuries per 100,000 does not total 100%. It is interesting that while permanent partial injuries account for only 5% of all injuries they account for 66% of the total costs.

Figure 3-7

4. The Key Advocate Axiom:

It is easier to persuade decision makers on matters of cost when at least one person within their own circle believes in the proposal well enough to champion the cause.

This is known as "lobbying" in political circles. Recognition of this axiom should be part of the planning strategy for any important presentation to "sell" a program or project which requires an investment of resources, such as, a "cost." Try to win at least one strong advocate who will support your proposal to a group or individual. The positive persuasion power of such a champion may make the difference between rejection or acceptance. While approaches will logically vary with each project or idea, a common one is to show the prospective champion how it will benefit him/her in particular, as well as others.

Damage/Waste Inventory

Area or Shop ________ Department ________ Supervisor ________ Department Head ________

1 Items Damaged/Wasted	2 Unit Costs	3 Estimated No. Damaged/Wasted Annually	4 Estimated Annual Costs	5 Actual Costs and Follow-Up Notes

The items determined to constitute the critical few, making up 80% of loss, when all items are evaluated will make the best targets for a loss improvement project team.

Figure 3-8

Damage/Waste Control Project General Information Form

1. Project No.	2. Subject	3. Report No.	4. Date
12	Grinder Damage	1	Feb. 15, 19–

5. Project Leader	6. Project Advisor	7. Frequency of Progress Reports
D. J. Brown	Ray Slider	Quarterly

8, Purpose of Project

To determine the cause of spindle fracture and stator burn-outs since the cost constitutes over 78% of grinding repair costs and is believed to be substantially outside the parameters of fair wear and tear.

9. Most Frequently Damaged Item or Injured Body Part	10. Main Accident Type Involved
Spindles and stators	Unknown
11. Departments or Areas Involved	**12. Supervisory, Staff or Technical Personnel Involved**
All grinding areas, Purchasing, Maintenance shop, Maintenance cost control	D. J. Brown, Paul Martin, B. J. Lowery, Park Dague
13. Equipment and Related Products Involved	**14. Occupations, Jobs and Crafts Involved**
B&D Portable Hand Grinder Model 20	Grinder
15. Maintenance Centers Involved	**16. Sources of Additional Information**
General Maintenance Shop	B&D Representative - R. Bunter

17. What Conditions or Practices Apparently Resulted in the Losses

Not accurately known at present. Based on early studies it is believed that striking the locking nut by the grinder in order to loosen it is causing crystallization that results in burn-outs since formal job observations apparently bring about substantial reductions during the periods of observations.

18. Project Objectives with Target Date for Progress

To reduce grinding maintenance cost 15% by July 25, 19– and 35% by December 25, 19– through engineering and/or grinding behavior controls.

19. Methods for Achieving Objectives

1. Design new wrench to provide proper torque on nut - May 15, target date. 2. Establish positive controls for operator grinding behavior - June 1, target date.

20. Additional Comments or Recommendations

The above methods for achieving objectives are based on early studies of this problem but are believed to possess excellent potential for control.

Figure 3-9

5. The Past Forecasts the Future Axiom:

Unless positive steps have been taken to correct or control past loss experience and if related circumstances continue to exist, you should expect to have similar if not more costly experience in the future.

We can study past economics involved with any area of a safety or loss control program and assume that, with no substantial change in those factors, future results will continue the trend. The value here is to use past cost information as a guide to action that will prevent history from repeating itself.

Perhaps one of the best applications of this axiom can be found in the insurance industry. Insurance companies use past loss experience as a prediction of future loss, and ultimately the basis for establishing their underwriting rates of coverage. These companies also furnish periodic computerized loss analyses to insureds, to indicate the critical few loss problems in terms of costs. Costs are, of course, based on recent past history and provide the insured with a vivid warning of future losses unless something is done to alter their occurrence.

Industrial engineers and efficiency experts use this axiom as the basis of much of their work in cost reduction. The past history of costs on machines or equipment, production maintenance, downtime and related damages are all used as a base line for many types of comparisons. The property damage and waste control material mentioned earlier, in the Axiom of the Critical Few, gives a detailed description of the "loss improvement project system" that makes extensive use of this axiom. Of course, those using this valuable axiom will seek adequate expertise to assure that their costing predictions are statistically accurate and a valid representation of what they are meant to communicate.

6. The Mutual Interest Axiom:

Projects and ideas involving costs are best sold when they bridge the wants and desires of both parties.

Supervisors/leaders who are best at selling projects and ideas involved with costs are those who clearly establish a bridge or connection of values between what "the company" wants and what you want. They seek out the benefits of ideas, projects and programs that contribute most to those items high on their motivation scale for the company.

Perhaps one of the best examples of applying this axiom can be found in Chapter 2 of this book, "*Safety Develops a Quality Prone Environment.*" To institute the objective method of housekeeping that insures the highest level of order does require significant training in a new concept. Time is costly, but justification is not terribly difficult when the program is properly presented to executive decision makers. The published testimony of quality assurance experts like Professor Robert Hayes of Harvard and Kiyoshi Suzaki, author of *The New Manufacturing Challenge*, will enable most executives to recognize that the benefits to the company in creating a quality prone environment far outweigh the relatively small cost of implementing the program. The truth of this axiom becomes music to the ears of those charged with making such cost benefit decisions. Isn't it easy to anticipate management reaction to statements such as the following by internationally known experts such as Mr. Suzaki?

> "We should understand the linkage between the level of housekeeping and the amount of defective products produced, the number of machine breakdowns, the routing of materials flow, inventory level, number of suggestions and the level of absenteeism."

Yes, when we bridge the wants and desires of both parties we improve our ability to present any new project, program or idea involving costs.

7. **The Economic Priority Axiom:**

   A manager will usually give priority response to items possessing the potential for the greatest proportion of results from the least investment of available resources.

This truth or principle is also expressed in a widely accepted economic corollary that a firm should choose from mutually exclusive cost control techniques the one which offers the highest rate of benefits to costs. It is difficult for some safety and loss control managers to understand why management will not financially support a specific program that promises a substantial potential return in cost reduction. Their understanding would be greatly enhanced if they were privileged to have all of management's information on both the financial resources available and all the goals and projects competing for those resources. One way to earn that privilege is to apply skills and techniques that lead to cost reduction through effective loss control management. This builds executive confidence in the individual and increases the probability that he/she will be more adequately informed on important aspects of the business.

It's ironic that we can understand why people purchase things for their family or house that give the greatest return for the money invested, yet find difficulty in understanding parallel management decisions. Like the average individual, management wants the greatest return for the limited resources available at any given time.

Hugh M. Douglas, co-author of the book, *Total Environmental Control*, and former Loss Control Coordinator for the Imperial Oil Company of Canada, made a statement that highlights the wisdom of this management truth:

We can take care of our people when we take better care of the business.

By utilizing available resources properly, management has more resources to utilize and, in effect, more money to spend on the needs of people and their safety and health. In the final analysis, this type of decision-making is absolutely necessary to assure the continuing existence of the business enterprise.

8. **The Vested Interest Axiom:**

Manager's are predominantly interested in those economic considerations affecting their own budgets.

One has only to ask managers a few simple questions regarding economics, to quickly determine where their interests lie. They are surely interested in the operating costs of the company, their own division, and to some degree, those of the employment or medical departments. But one thing will be made crystal clear in the responses — they are predominantly interested in the budget they are accountable to achieve, and by comparison, only slightly interested in anyone else's. This deep, vested, personal interest is very logical when we consider that promotions, merit increases and other things that really count, largely depend on one's ability to contribute to the profitability of the organization. It is one's own budget, of course, that best reflects the management skills that count in this regard. Likewise, a supervisory career can be short lived, if that person consistently fails to manage the controllable costs. Through the years, failure to consider this principle in the application of costs with management people has been a major shortcoming of many safety/loss control managers.

While terms like "insured - uninsured" costs can have significant motivational value when properly applied, they also can become sources of subtle resistance and underbreath jokes when used inappropriately. It is absolutely naive to believe there is any significant motivational value in using costs related to the medical, employment or safety departments with an operating manager. There are significant uninsured costs that have direct bearing on the manager's own budget, but in many cases he/she is not aware of the impact they have on the attainment of his/her budgetary objectives. Results of accidents and other undesired events involving property damage, waste, overtime, rework and delays are buried in costing categories that camouflage their controllability. For instance, property damage costs are usually buried in a general maintenance category.

For similar reasons, terms like "direct" and "indirect" have become less and less part of the professional's vocabulary. The word "indirect" not only has been associated with many items external and remotely concerned with the individual manager, but also bears the inference of "remoteness" — something that would be difficult to do anything about. For these reasons, there is an increased trend to use the terms "ledger" and "non-ledger" costs. The term "ledger" carries the strong personal budget implications of vested interest for the manager.

It must be remembered that the budget of a plant manager is a composite of many smaller budgets of division managers, department heads and front-line supervisors. That plant manager may very well have a vested interest in a broader variety of uninsured costs than one of the subordinates. Regardless of the level of the manager involved, the principle remains the same. Cost items are more relevant when they can be associated with the budget the particular individual is accountable to manage. One of the greatest challenges facing safety or loss control coordinators is to develop techniques and skills which enable them to clearly identify specific items of loss that relate in a meaningful way to the manager involved.

9. **The Axiom of Adequate Information:**

   The timeliness of managers' decision-making is directly related to the adequacy of information they have upon which to act.

Through the years, one of the most common outcries of safety, environmental health, security and fire protection managers alike, has been directed toward failure of management to act promptly on decisions involved with hazard correction. While there may sometimes be justification for these concerns, on many occasions there is also much to be said in defense of the delayed management action. It is not difficult to understand the frustration of a program coordinator who has organized a planned inspection program that provides detailed reports on all plant areas month after month, only to have them seemingly ignored. Neither is it difficult to understand that a busy manager has many important things to do, in addition to taking the time to read voluminous reports that require interpretation and organization of thoughts in order to give intelligent direction. These kinds of frustrations and problems for both staff and operating managers will become less the rule and more the exception as managers in safety and other loss control disciplines learn to manage more professionally. Providing management with an increased number of devices and techniques to aid decision-making will bring the application of loss control in general industry more in tune with the technological advances of our information age.

The inspection form shown in *Figure 3-10* reveals how hazard classification and the use of symbols can save busy managers time in screening inspection reports, assist them in making spending decisions and provide ready guidance on the priority of attention they should give to problems.

INSPECTION REPORT				Department Machine Shop
Inspector(s) Bob Mettner, Paula St. John			Area Inspected Plant #2	Reviewer John Sharp
Item Number	Hazard Class	Type of Report (Initial/Follow-Up/Final) ~~Initial~~ Final	Date(s) 8/13/19– 8/31/19–	Quality Score 86%
		ITEMS DETECTED - ACTIONS TAKEN - DATES		
*1	A	6/10/19–, guard missing, cutting head machine #2046. NE corner Bay 2/B		
*2	C	6/16/19–, door at S end bldg. warped, hard to open. WO issued to		
		carpenter shop, work scheduled for 7-22.		
*3	B	7/8/19–, heavy accumulation of oil and trash under main motor in bldg.2		
		pump house. Cleaned out Aug. 20, discussion with workers all turns,		
		permanent signs posted.		
4	B	Two pallets of chemical 265 in yard, column B14. Directions warn against		
		outside storage. Reported to R. Jones, area supervisor. Pallets moved to		
		inside storage Aug. 6.		
5	A	Major building column P32 is receiving severe damage from bulldozer		
		operation cellar bldg. 4. WO issued for concrete barricade base Aug.12,		
		meeting held with all operators on major damage.		
6	A	Seven leaks in natural gas line between columns A1 and A2 in N bay,		
		bldg. 2. Strong odor of gas in entire area. G. Sutter, area foreman		
		roped area off. Gas dept. called. Leaks repaired Aug. 6.		
7	C	Sharp edge on latch - tool shanty door bldg. 2. Maintenance worker		
		filed surface Aug. 7.		
8	B	Compressed air pipe N side of truck door, E end of shop damaged. Guard		
		or relocation could prevent major loss. WO issued 8-9 for permanent		
		striped metal guard - hazard striped board wired to pipe Aug. 10.		
9		Two empty whiskey bottles found in empty locker #15 - washroom bldg. 1.		
		Plant protection notified Aug. 7, request area sup. to permit		
		observation before discussing.		

Figure 3-10

The A.B.C. classification indicates the severity of the hazard. An asterisk(*) indicates an item carried over from a previous report, and a circled number tells the busy manager that intermediate action has been taken on the particular item. An item crossed out (x) tells the reader that correction has been completed.

More and more organizations are including a small, yet very important, section on supervisory investigation forms to quickly guide the supervisor on how much time to spend on his remedial action, and give inference to costs that can be justified for control of future events. (See items 23 and 24 on *Figure 3-11*.)

One of the more important requests that come to safety and loss control coordinators from operating managers is for help on evaluating potentially hazardous conditions and proposed solutions. *Figure 3-12* presents five questions that a staff professional or operating manager can use to assist in evaluating a potential or real hazard, deciding on the need for action, and justifying the money for its remedy. Note that evaluation can be in the form of terms or words, letters, numbers or some combination. The important thing is that management can be given valuable information on all critical aspects of hazard evaluation, to help them give a timely response regarding cost justification and priority. The Guide to Risk Decisions (*Figure 3-12)* is a very practical tool for this purpose, especially when significant expenditures are involved.

An increasing number of companies are classifying accidents, as well has hazards, in their loss control programs. The classification is generally based on the degree of liability anticipated, and its related economic implications. The enormous court awards involved with product and general liability cases require that managers consider their involve-

Investigation Report

IDENTIFYING INFORMATION

1. Company or Division			2. Department	
Western Box Company			Sheet and Paint	
3. Location of Incident		4. Date of Incident	5. Time AM PM	6. Date of Report
#2 Shipping Dept. S Wall Col 20B		4/18/19–	3:20	4/18/19—
Incident or Illness		Property Damage		Other Loss
7. Injured's Name		14. Property Damage		15. Type
Robert K. Berry		Three machine guards		
8. Part of Body	9. Days Lost	15. Nature of Damage		19. Cost
upper left leg	13	Badly bent		
10. Nature of Injury or Illness		16. Cost Estimated	Actual	20. Nature of Loss
Compound fracture		125.00	110.00	
11. Object/Equipment/Substance Inflicting Harm		17. Object/Equip/Substance Inflicting Damage		Obj/Eq/Sub Related
Misc. parts on floor		Worker falling on parts		
12. Occupation	13. Experience	22. Person in Control of Activity at Time of Occurrence		
Painter	8 years	Robert K. Berry		

RISK

Evaluation of Loss Potential If Not Corrected

23. Loss Severity Potential
☐ Major ☐ Serious ☐ Minor

24. Probability of Occurrence
☐ High ☐ Moderate ☐ Low

DESCRIPTION

25. Describe How the Event Occurred

Bob was reaching out with a paint brush while standing on the third rung from the top of a 20′ ladder section, the top section of a 40′ extension ladder. He was attempting to paint a small unpainted area of the oxygen pipe running along the south wall of the #2 shipping department. He lost his balance as the bottom of the ladder started slipping and fell approximately 10′ to the top of a pile of miscellaneous parts and machine guards piled on the floor.

CAUSE ANALYSIS

26. Immediate Causes: What Substandard Actions and Conditions Caused or Could Cause the Event?

Attempting to reach too far. Ladder positioned at too great an angle, probably due to disorderly storage under the pipe. The ladder did not have safety shoes and was not tied off to the pipe since the tie-off rope was missing. Also, Bob was not using his safety belt.

27. Basic Causes: What Specific Personal or Job Factors Caused or Could Cause This Event? Check on Back, Explain Here.

A check indicates that Bob's last review of ladder rules and standard practices was over 3 years ago. There are no recorded job observations in Bob's file and his record is free of disciplinary actions. Ladder inspections have been informal in the past unless picked up in general inspections. The ladder section was not stenciled with the required warning against using it alone.

ACTION PLAN

28. Remedial Actions: What Has and/or Should Be Done to Control the Causes Listed?

1. Rule and standard practice review with painters will be completed within 30 days. 2. All ladders were inspected immediately after the accident and a monthly ladder inspection will be made and recorded. 3. A paste-in rule is being prepared requesting painters to inspect their equipment daily before use and to date/sign the tab. 4. Department meeting with Carl Doan to institute a planned job observation program. 5. A separate report on housekeeping is being prepared.

Figure 3-11

ment in accident investigation as a critical part of their loss control effort. Accident classification generally guides managers in determining the depth of the investigation necessary, depending on the potential severity of the loss. One utility company uses the classifications shown in *Figure 3-13.*

Guide to Risk Decisions
Key Questions

CLASSIFICATION OF EXPOSURE

1. What is the potential severity of loss if an incident occurs?
   A - Major
   B - Serious
   C - Minor

PROBABILITY OF OCCURRENCE

2. What is the probability that a loss will occur from this exposure or hazard?
   A - High
   B - Moderate
   C - Low

COST OF CONTROL

3. What is the cost of the recommended control:
   A - High (establish meaningful cost ranges
   B - Medium for your own organization)
   C - Low

DEGREE OF CONTROL

4. What degree of control will be achieved by this expenditure?
   A - Substantial or Complete (67-100%)
   B - Moderate (34-66%)
   C - Low (1-33%)

ALTERNATIVES

5. What are the alternative controls?

JUSTIFICATION

6. Why is this one suggested?

Figure 3-12

Critical	Personal Injuries	Property Damage
Class 1	More than one public (non-employee) death.	$500,000 +
Class 2	One public (non-employee) death or hospitalization of five or more public.	$250,000 to $499,999
Class 3	More than one employee on-the-job death.	$100,000 to $249,999
Class 4	Hospitalization of two or more public or one employee. Death or hospitalization of five or more employees (OSHA report).	$50,000 to $99,999
Major	Disability (loss time) of one-four employees or of one public (non-employee).	$1,000 to $49,999
Recordable	Workers' Compensation Medical Care case.	$0 to $999

Figure 3-13

Loss control specialists in petrochemical operations regularly employ sophisticated techniques to mathematically evaluate risks and determine the potential losses, should the undesired event occur. These risk analysis techniques establish cost factors such as the maximum probable loss exposure; the maximum possible loss that might occur from the foreseeable event, with protection facilities not functioning; and the maximum probable loss to be expected when control function as intended to limit the loss. While fire protection engineers representing insurance underwriters apply these techniques in general industry, there is a great need for local program coordinators to expand their use, and provide management with additional decision-making tools.

Another example of a tool to provide management with more adequate information upon which to act can be found in the suggested form shown in *Figure 3-14.* It is extremely difficult for managers to evaluate the hazard involved, as well as the economic value of a safety or health related suggestion, when adequate information is lacking. A tool like this gives more adequate information.

10. The Axiom of Dimensional Value:

The degree of management attention is directly related to the size of the problem.

A classic story is told of the corporate safety manager who went to his executive vice president with plans to produce a motion picture on preventing finger and hand injuries. The executive asked what the cost of the film would be and what reduction of hand injury costs was anticipated by the proposed expenditure. When told that the film would cost $40,000, that the costs of finger and hand injuries exceeded $50,000 during the past year, and that any reduction in those costs was difficult to determine, the executive's response was most interesting. "Jack," he said, pulling his operating budget from his middle desk drawer, "we paid $55,000 last year for toilet paper, and my manager of services tells me that by going to a 3 1/2" roll rather than the standard 4 1/2" roll we presently use, we can immediately cut our annual costs $10,000." Needless to say, this safety manager was made aware that this executive was not impressed with either the dimension of the total loss figure or the uncertain plan to reduce it.

Safety and loss control managers have recognized for many years that medical and compensation costs are not usually sufficient to provide motivation their programs deserve. To this day, many safety people use H.W. Heinrich's 4:1 cost ratio (the total cost of an injury is four times the medical

Loss Control Department
Suggestion Investigation Report

Suggestion Number:	Date:

Details of Investigation are as follows:

☐ Not original ☐ Routine maintenance ☐ Not suggestible ☐ Covered by former suggestion ☐ Violation of safety rule ☐ Present standard practice violation ☐ Cost too great for expected results ☐ No hazard with normal caution ☐ Should become standard practice for same type items in future		1. Degree of hazard: ___Minor ___Major ___Extreme 2. Frequency of exposure: ___Light ___Medium ___High 3. Extent of Application ___Local ___Restricted ___General 4. Degree of Elimination: ___Slight ___Substantial ___Complete 5. Originality: ___Not new ___Partly new ___Entirely new

Recommendation

Acceptance:	Rejection:	Trial:	Additional Info requested
Investigation made by:		Discussed with:	Approved:

ALL SUGGESTIONS must be answered in detail promptly following a thorough investigation at the location concerned. The investigation should always include thorough questioning and evaluation of the opinions of persons involved and should thoroughly answer the what, why, when, how and who of all the suggestion aspects.

Figure 3-14

and compensation costs), in an effort to show management a substantial enough problem to gain additional attention. Many colleges and universities conducting graduate programs in safety studies give considerable attention to the insured-uninsured cost concept to equip future safety and loss control managers with the ability to demonstrate the magnitude of the accident loss problem.

The authors certainly concur with the generally accepted premise that the use of costs can add a significant motivational dimension to a program. They also believe deeply that medical and compensation costs alone have not generally provided a large enough base when compared to the operating costs of a typical company. While it is entirely possible that the fast-rising insurance and compensation costs may change the exception, to be the rule, this is not probable for some time. This does not mean that management is uninterested in on-the-job injury costs. They are interested in all operating costs, especially during a period of inflation and reduced business activity that so adversely affects profitability. What the axiom of dimensional value tells us is that we do not gain the necessary management attention, unless we present the costs of our losses in their true magnitude and economic perspective.

Check the lists of cost items related to typical safety programs to determine how many of them are accurately known and utilized properly in your own motivation efforts. A glance at the estimated annual accidental losses to industry (*Figure 3-15)* shows the magnitude of the *total* accident cost problem compared to that of the on-the-job injury area alone.

We believe that the reason that over 95% of the vice presidents of safety and loss control are in motor fleet and railroad operations is directly related to their application of

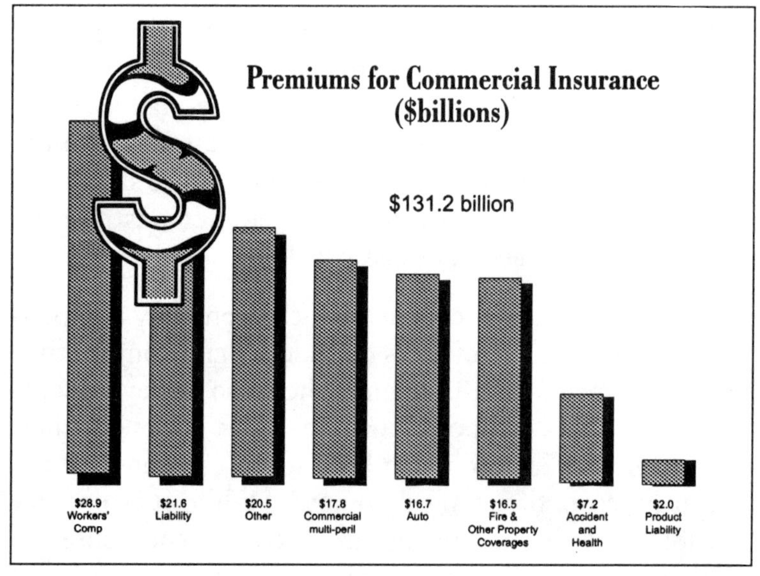

Figure 3-15

this principle of dimensional value. They precisely identify accident costs involved with injury, property damage and liability, and direct well organized programs at these enormous targets. The dimension of their cost targets is large enough to gain significant management interest and attention. Their personal recognition is further evidence of the value that management places on their effort. While we use the safety discipline to illustrate this point, the coordinators of security, fire, environmental health and other individual loss-related programs should recognize that this principle has identical application to their areas. It is also our opinion that an excellent career development path for a specialist in any one of these fields is toward loss control management. Precedence for this can be found in the petrochemical industry in particular.

The following charts present these two classes of insurance that can best be estimated. However, some forms of insurance often purchased by individuals, such as farm owners' insurance or coverage on high-value personal property, are normally included in commercial insurance. Thus, these estimates are not precise. Note that these figures are based on the net written premiums. They were obtained from *The Fact Book: 1996 Property/Casualty Insurance Facts* published by the Insurance Information Institute of America. These costs should be considered the very minimum costs of risk since they do not include costs below or above insurance coverages, general property damage (which has been reported by numerous sources to be five to fifty times the cost of workers compensation) and the costs incurred by those totally or partially self-insured.

While the potential today for a multi-million dollar loss from a single event, such as the death of a young outside contractor with a large family, is quite realistic. It will be some time before these individual risks provide the motivational force for management, that their everyday common losses provide. When we combine the everyday costs involved with the interrelated disciplines of safety, environmental health, fire and security, we then possess a package of cost reduction potential of unquestionable motivational dimension.

The persuasive value of using costs that more truly represent the impact on general operating costs was made quite clear in a recent talk by Raymond H. Marks, former President of Tenneco Chemicals, Inc. Mr. Marks made the following comments:

> For purposes of understanding, let me state that a sound loss control program is not limited to employee safety and health — although this is an important segment and our first consideration. A comprehensive loss control program is aimed at elimination of problems of occupational health,

> property protection, product safety, security, and any other area where an unintended incident can occur and detract from the company's profitability.... Let me assure you that by taking this broader view of the problem, we in no way diminish our emphasis on protecting the safety and health of our employees. To the contrary, emphasis on a broader range of subjects creates a continuing greater awareness and complements our efforts to protect individual employee safety and health.... Safety and loss control are an important part of the executive suite. No longer second-class citizens, loss prevention and profit performance have become synonymous. There is no room for compromise. Total dedication and re-dedication are the key ingredients.... How often do we get the chance to demonstrate our innate concern for the health and well-being of our fellow humans, while at the same time, we improve our profit performance. Think about it_total loss control programming is sound business planning, the best of both worlds under one roof.

Perhaps there is no greater way to illustrate the cost reduction side of the double-edged sword referred to by Mr. Marks, than to consider the sales volume required (see *Figure 3*-5, page 88) to make up the profit lost by the undesired events in the absence of a good loss control program .)

CONCLUSION

These, then, are ten axioms of economic application that can be utilized to motivate increased management interest and action in a loss control program. You can use them individually, or in combination; but most important, use

Safety Applications of Economic Axioms

1. The management control axiom.
2. The axiom of economic association.
3. The axiom of the critical few.
4. The key advocate axiom.
5. The past forecasts the future axiom.
6. The mutual interest axiom.
7. The economic priority axiom.
8. The vested interest axiom.
9. The axiom of adequate evidence.
10. The axiom of dimensional value.

them. While there is no doubt that their effective application will require an enormous amount of innovative application of our best professional efforts and energy, the rewards that will come with success will be more than justified. We have everything to gain including the management commitment we must have for success.

Selected References

Accident Facts. 1995 Edition. Itasco, IL: National Safety Council, 1995.

Accident Prevention Advisory Unit. London: Health and Safety Executive, 1994.

Bird, Frank E., Jr. and George L. Germain. *Practical Loss Control Leadership.* 4th Edition. Loganville, GA: Det Norske Veritas, 1996.

Bird, Frank E., Jr. and Robert G. Loftus. *Loss Control Management.* Loganville, GA: International Loss Control Institute, 1976.

Grimaldi, John V. and Rollin H. Simonds. *Safety Management.* 5th Edition. Boston: Irwin, 1989.

Workers' Compensation Claim Characteristics. Hoboken, NJ: National Council on Compensation Insurance, 1995.

Property Casualty Insurance Facts. Insurance Information Institute, 1995.

Suzaki, Kiyoshi. *The New Manufacturing Challenge.* New York: The Free Press, 1987.

The Cost of Risk Survey. Risk & Insurance Management Society, Inc. & Tillinghast, a Powers Persin Company, 1994.

Chapter 4

A Quality Expert's View

INTRODUCTION

A long-standing belief among quality professionals is that quality is a unique and identifiable profession based upon a defined set of management technologies. It is a business process that has developed and found its place in the management framework of organizations, just as H-S-E (Health - Safety - Environment) surely will. Quality has grown from its early roots of artisan inspections of products to become a critical theme in corporate management strategy. It is amazingly similar to the evolution of H-S-E, from inspection to management. This raises several areas to consider when thinking about the role quality plays in an organization.

First, it has developed into a series of philosophies, technologies, methods, and customer expectations that have helped advance industry, service, and government to new levels of performance. It has had a very significant boost by a series of historical and market shifts that H-S-E has not enjoyed. Its evolution of these practices has stimulated higher levels of product utility, efficiency, effectiveness, value, and cost. One only has to drive a 1950's vintage

This chapter contributed by Stanley G. David, Director, Management Consulting & Rating Services, DNV-Loss Control, Houston, TX.

automobile to get a perspective on how far we have come. Of course, many EHS leaders are still searching for ways to gain greater management commitment. While still seeking continuous improvement, there is no doubt that great progress has been made through the years.

As the sophistication of the products and customer's values improved, quality techniques became more sophisticated as well. (Factors such as aerospace safety and OSHA requirements have had similar developmental effects on safety.) These developments have created a quality industry that applies these methods to business or service problems similar to a physician performing surgery to correct a medical ailment. Like the patient, the company is somewhat ignorant of the range of possible treatments and must rely on the professional to properly select and apply the treatment. Not treating the ailment, or relying on self-inflicted cures, sometimes leads to undesirable results. Therefore, organizations are under a reasonable amount of pressure to adopt and employ quality in order to stay "fit" enough to keep up with global competitors who use all of the quality tools available, such as ISO 9000. They use quality not only for resolution of problems, but also for prevention of problems and ultimately the extension of performance. There is a substantial indication that safety will follow a similar path by way of ISO standards.

Second, quality has become a market imperative. The global acceptance of ISO 9000 and its offspring has revolutionized global quality values (European leaders strongly believe

EHS will not be far behind). Virtually every developed nation has voluntarily adopted a set of standards for quality system management. This system has stimulated tens of thousands of firms to improve their process of under-

This chapter contributed by Stanley G. David, Director, Management Consulting & Rating Services, DNV-Loss Control, Houston, TX.

standing their customer's requirements, then putting in the infrastructure to deliver products that meet them consistently. Customers will be the ultimate beneficiaries of this marketplace evolution. Its full effects on industry, service, and government are unknown. Whatever the outcome, a vast number of businesses, services, and government agencies will be employing formalized quality methods as a part of their day to day operations. Those close to the changes taking place see EHS following a parallel acceptance route. Some European countries are demonstrating the synergy between quality and safety, as indicated in Chapter 2. These efforts will certainly aid substantially in gaining the commitment required for safety excellence.

Finally, as a result of this increased management interest in both quality and safety, organizations are being asked to address other societal values beyond customer satisfaction and short-term profitability. Issues such as environmental protection, public health, community service, wellness and ergonomics have the potential of impacting organizations and the world in much the same way quality has. This chapter will explore some of these similarities, and as with brothers, some of the differences.

> In many ways EHS and quality are like siblings. They have common genes and mannerisms. They have similar work processes. They have a common set of values pointed at improvement.

WHAT IS QUALITY?

Quality has so many meanings to people that one can become confused whenever this term is used. The Malcolm Baldrige National Quality Award defines quality as:

Quality is judged by customers. All product and service characteristics

This chapter contributed by Stanley G. David, Director, Management Consulting & Rating Services, DNV-Loss Control, Houston, TX.

that contribute value to customers and lead to customer satisfaction and preference must be a key focus of a company's management system.... This concept of quality includes not only the product and service characteristics that meet basic customer requirements, but it also includes those characteristics that enhance them and differentiate them from competing offerings.

More simply, quality is a set of philosophies and disciplines focused upon the principles of improvement, waste reduction, customer satisfaction, and market differentiation. Quality assurance and control has gone through three definable generations:

- *Generation 1 — Inspect out defects*
- *Generation 2 — Engineer out defects*
- *Generation 3 — Variance free system*

Firms tend to evolve from one generation to the next and some firms have elements of all three in their quality programs. The purpose of a formal quality program is to provide the specialized service, knowledge, and discipline needed to support the quality strategy of the organization. Some organizations have traditional quality assurance and quality control programs that utilize quality engineers, inspectors, test technicians, and other specialized administrative personnel that are interwoven into the production process. Other organizations opt for a facilitated approach. Normal operating personnel are expected to perform the tasks that inspectors or specialized quality personnel might otherwise perform. In this situation, quality personnel serve as teachers, auditors, coaches, and technical resources, functions like those served by safety professionals. The range of services provided by quality personnel in a quality program can be quite broad:

- Product Inspection, Testing, Evaluation and Certification

This chapter contributed by Stanley G. David, Director, Management Consulting & Rating Services, DNV-Loss Control, Houston, TX.

- Process Analysis and Statistical Evaluations Used to Identify and Solve Problems
- System Auditing
- Nonconforming Material Disposition, Control, and Analysis
- Quality Performance Measurement, and so forth.
- Selection, Development, and Use of Sophisticated Measuring Equipment
- Quality Planning
- Customer Quality Liaison
- Customer Complaint Handling
- Quality System Design, Implementation, and Management

These roles and missions are carried out by various categories of personnel who vary considerably depending on the level of maturity, sophistication, and size of the firm's quality program. In some firms these roles are considered as a part of the value adding processes since the customers feel it is of value as well. For example, a large jet engine manufacturer has a sophisticated series of tests that are performed to establish that a new engine performs as planned throughout its full range of operating parameters. These tests are mandated to certify the safety and performance of the engine to the customer and to federal aviation authorities before it ever gets near a passenger carrying airliner. This certification activity requires very expensive test cells, sensory equipment, support services, and highly trained personnel. These requirements follow a pattern quite similar to those of safety in similar circumstances.

In less complex and non-regulated product lines the quality regimens may be purely voluntary on the part of the product provider. For example, an auto tool supplier may only do

This chapter contributed by Stanley G. David, Director, Management Consulting & Rating Services, DNV-Loss Control, Houston, TX.

cursory checks on finished products since the customer's expectations are not very demanding for this product line if the tools do their intended function. In this environment specialized quality personnel and practices may not be available or widely employed. These examples represent some of the following dimensions of a quality program. This may be one of the few big differences between application of certain quality measures and safety. In addition to the demands of more hazardous conditions, certain circumstances may be adjusted to comply with OSHA while the quality program is predominately customer driven.

DIMENSIONS OF A QUALITY PROGRAM

To more fully understand the breadth of similarities between methodologies like quality and safety management, it is useful to break quality into component parts and look at the major elements in greater detail. The balance of this section is devoted to this analysis. An overview of each element from a quality management perspective will be reviewed to provide a comparison with safety management.

Continuous Improvement & Improvement Strategy

The process of continuous improvement is based on the concept that employees, regardless of their position, should endeavor to identify and improve their work every day. This process, in practice, ranges from very informal instructions to employees, to formally managed programs. Milliken & Company is a good example of a formalized program. It uses an Opportunity For Improvement (OFI) form to document all new improvement ideas. These OFI's are then carried through a planned process that evaluates, judges, and launches accepted OFI's for action. Milliken managers are evaluated by the number of OFI's that they successfully implement each year.

This chapter contributed by Stanley G. David, Director, Management Consulting & Rating Services, DNV-Loss Control, Houston, TX.

Continuous improvement is a key area of focus for most quality methodologies. In theory, if all associates focus on improvement of all aspects of their work on a continuous basis, the constant stream of improvements will result that will improve business performance. This is very much the same direction often given to EHS programs. The focus on a never ending stream of continuous improvements is a logical step that most managers tend to endorse as a part of their management system. Many of the classical quality philosophies espoused by Deming, Juran, Crosby and so forth, rely heavily on continuous improvement as the way to reduce costs and improve profits.

As with other philosophies, continuous improvement has some pitfalls as well as promise. The greatest of these pitfalls is its lack of focus and uncontrolled use of resources. Many companies who have used continuous improvement for years became disenchanted when they did not see bottom line improvements commensurate with the effort expended. The reason is simple. If each person in a company invests time and money on improving things that they perceive as important, we get a wide range of results. Examples are:

1. An employee finds a simple area of waste and corrects the situation saving time and money.
2. An employee team attacks a chronic problem and develops some new methods that eliminate the problem with a minor investment on the part of the firm.
3. An employee does not clearly understand underlying business issues and is confounded when efforts to improve a process are blocked by larger business priorities outside his/her span of control. This person spent many weeks of effort before being asked to stop by management.

This chapter contributed by Stanley G. David, Director, Management Consulting & Rating Services, DNV-Loss Control, Houston, TX.

4. A quality circle spends months improving the performance of a work cell and, after investing in new tooling and methods, reduces total process cycle times. However the shop is already operating below capacity and the 'improvements' further exaggerate the situation by creating more capacity. Since the market is flat, associates are laid off after the firm decides to outsource the work and divest the manufacturing cell.

The above examples are real world cases and they do highlight some of the limitations within the continuous improvement philosophy. *Continuous improvement cannot be a stand alone effort.* It must be tempered and lead by management, then used where appropriate. Having associates spend a lot of their time and energy improving something that does not make strategic sense for the business is not only illogical, it is wrong. It creates a lot of bad feelings on the part of the associates who see the results of their efforts being capsized by business priorities. It is therefore important to align these improvement efforts with the goals of the business. The most common way is to have continuous improvement projects focus on a set of company priorities that are communicated to the associates. This same sort of focus is increasingly becoming an established part of safety programs that are showing strong measurable results.

These priorities tie to the business and financial strategy, so efforts made on the part of the associates to improve performance will be more likely to support the true needs of the business. A classic example of a company who unleashed its associates to improve everything imaginable with hope that results would follow is the Wallace Company. Even the Malcolm Baldrige National Quality Award did not stop the firm from entering bankruptcy when they lost financial focus on expenses and cash flow. Since then, the company has focused its improvements on cash flow and is

This chapter contributed by Stanley G. David, Director, Management Consulting & Rating Services, DNV-Loss Control, Houston, TX.

now recovering. (The "critical few" approach discussed in Chapter 3 applies to quality as well as safety).

Therefore, the most important factors in the selection of improvement projects are the strategic requirements of the business. In order to align business and quality priorities, management must consider the financial and non-financial goals of the strategic plan when choosing improvement priorities. It must determine the competitive impact of improvement priorities and it must determine the competitive impact of potential projects.

> Some simple truths have been discovered: (1) to achieve competitive success, the selection of significant projects is much more important than the total number of improvement projects that are activated, and (2) it is essential that everyone involved in the selection process understand the achievement of specific strategic and financial results will only occur if improvement project selection is based on logical cause-and-effect relationships and not pure faith in continuous improvement dogma.

The strategic requirements of the business determine whether a project is important or not.

To establish the infrastructure for effectively managing improvement on a continuous basis the firm needs:

1. Good mechanisms to communicate financial and business strategic needs to the associates so that they can contribute to the "right stuff".

2. A process to evaluate and select improvement projects which ensures that the energy to be focused on improvement is appropriate and timely with respect to business imperatives.

3. A mechanism to record improvement ideas and track

This chapter contributed by Stanley G. David, Director, Management Consulting & Rating Services, DNV-Loss Control, Houston, TX.

them. Even if an idea is not selected for immediate action, if it has merit it can be placed in an "Idea Bank" for future action as resources and priorities allow.

4. Training for associates to give them the tools and confidence to confront change and make improvements.

5. Feedback and reinforcement mechanisms to associates who contribute to improvements so that they know their efforts are noticed and appreciated.

If the above are applied consistently, evidence suggests that continuous improvement does contribute to improved business performance.

This discussion of continuous improvement applies to the EHS realm as well. The same rules apply for the effective implementation of continuous improvement relative to Safety and/or Environment. We know there are thousands of companies doing it .... but thousands more that should.

Most likely the biggest reason there is not the same level of continuous improvement in EHS as in quality is the lack of commitment that the customer-driven quality emphasis has achieved. When ISO introduces standards for safety as they have for quality, we will see the influence of European import requirements on international customers. In time, the same customer-driven performance requirements will stimulate continuos improvements for EHS as for quality, for those organizations desiring international trade.

EHS improvement programs can then be comfortably merged into quality improvement programs, or vice versa, or into a company improvement program that does not differentiate between the different areas. The key to this sort of approach is not to lose focus on any of the constituent areas.

This chapter contributed by Stanley G. David, Director, Management Consulting & Rating Services, DNV-Loss Control, Houston, TX.

Customer Requirements

In the quality system, "customer requirements" is the typical purpose for being. The primary metric tends to be some form of customer satisfaction measurement against specified customer requirements. Some quality leaders and systems such as ISO 9000 have gone so far as to define quality as the "meeting of customer expectations and requirements." These systems require that infrastructure to ensure customer requirements are defined, properly interpreted, and achieved.

Customer requirements can vary:

- Specific technical requirements such as output voltage or viscosity
- Performance requirements such as miles per gallon or range of transmission
- Visual or sensual requirements such as what color or how soft
- System requirements such as compatibility with television broadcasting formats
- Certification requirements such as Underwriters Laboratory or ISO 9000
- Documentation and data requirements
- Engineering requirements
- Delivery requirements
- And many others.

The typical quality system is designed to help the organization manage and meet the above types of requirements. Moreover, the infrastructure is designed to allow the organization to respond to new and changing requirements as well. In most cases, it does not matter what the requirements are, the systems are designed to help ensure that they are dealt with.

This chapter contributed by Stanley G. David, Director, Management Consulting & Rating Services, DNV-Loss Control, Houston, TX.

The few cases where the quality system has not fully addressed requirements are in the areas of safety and environment. Possibly because these areas are typically communicated through different channels, these areas are often left outside the "scope" of the quality system. Moreover, since standards like ISO 9000 have ignored the safety and environmental disciplines completely or in part, many times the company's quality systems do as well.

With the advent of ISO 14000 this may change, for environmental management. An ISO safety standard is strongly rumored as well. If these systems do come into being, and are widely adopted, then a truly common system for identifying and managing customer requirements would be in force.

Do customers require safety systems? Yes, sometimes directly, as in the case of product safety, and often indirectly. Some customer contracts have safety provisions or requirements built in. In those cases, the organization's responsibilities are clear. In other cases, the requirements may be inferred or assumed as in the firm's meeting of applicable safety laws such as OSHA.

The management system infrastructure that addresses technical and quality requirements should be designed to identify and address safety requirements as a part of normal practices. The anticipated ISO standards may bring this about sooner than many anticipate.

Cycle Time Reduction

One of the more popular and productive elements of modern total quality management programs is the focus on reduction of process cycle times. Gurus such as Philip R. Thomas have done a lot to catch the interest of industry. The global competition of reducing cycle time in industry has turned into a mania. The simple strategy behind cycle

This chapter contributed by Stanley G. David, Director, Management Consulting & Rating Services, DNV-Loss Control, Houston, TX.

time reduction is to cause large reductions in the total time it takes to design, make, deliver, or serve products by making fundamental changes in the way work processes are designed.

An example is using the Thomas strategy to slash inventory and lot sizes while drastically shortening set-up costs and times so that jobs can be completed in days rather than months. If successful this reduces the need for sophisticated inventory and production planning systems plus the complex infrastructure needed to control them. The organization is then handling a relatively small number of jobs at any one time since cycle times are so much shorter. This task of getting from a slow, inventory based, large scale producer to becoming an agile manufacturer is difficult but rewarding. If the management team can learn to adapt to the new order of business the firm can operate at much lower cost and overhead levels while producing huge increases in volume with great flexibility. Since this equates to profit, cycle time reduction is now a fairly well accepted goal in today's boardrooms.

Inventory Covers Problems

In order to accomplish these gains the firm faces the well-known "alligator pond." That is as the firm drains inventory, which has a tendency to mask problems, business cycles speed up and hidden problems surface with visible impact. The firm quickly must face and eliminate "alligators." In his series of books Eliyahu M. Goldratt refers to these problems as constraints that have to be eliminated. Management's role thus changes from building enough inventory to mask inefficiencies, to spending their time eliminating problems that constrain the performance of the business. In theory the business gets progressively more efficient and simpler to manage.

This chapter contributed by Stanley G. David, Director, Management Consulting & Rating Services, DNV-Loss Control, Houston, TX.

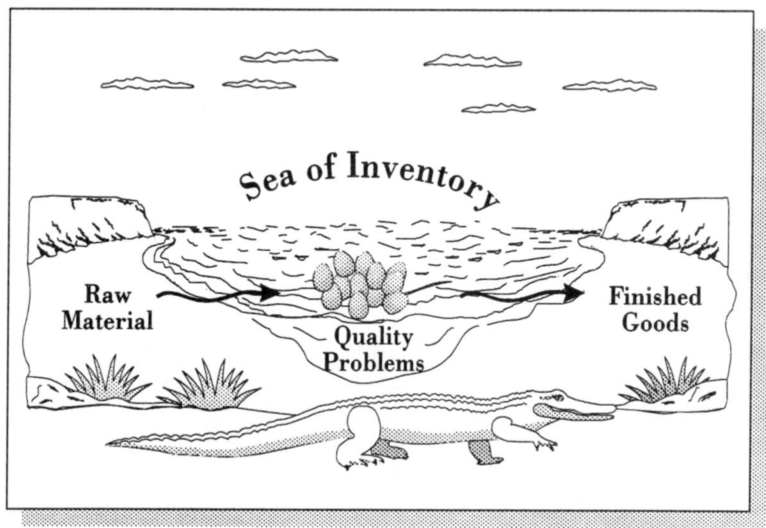

Thousands of organizations are going through this process with many documented success stories. Others have failed due to cultural considerations that would not allow the owners, managers, employees, unions, or a combination of all of these to make the huge personal and business changes needed to succeed. Wasteful production practices and non-conforming materials breed problems that can result in unanticipated events that do increase the risk of incidents as well as inefficiency. (see Chapter 2 in this regard).

> From the Safety perspective the benefits gained from cycle time reduction can also be large. As an organization clears bureaucracy and inefficiency from its processes many of the contributors to poor safety performance can be addressed as well. A well-organized agile manufacturer not only runs a fast and efficient shop, they run a safe one as well since the optimization of internal processes requires good process analysis disciplines that take human factors into consideration.

Shorter cycle times also lead to better organized shops. The

This chapter contributed by Stanley G. David, Director, Management Consulting & Rating Services, DNV-Loss Control, Houston, TX.

reduction in inventory and quicker processing of smaller lots tend to help clean up and organize the business. Since a lot of the problems that forced jobs to sit in place for a long time in the slower, inventory-based system are no longer tolerated in a short cycle time environment, a lot of work in process and its inherent problems no longer exist. The results tend to be better organization and cleaner production areas. This order and cleanliness have many benefits in reducing quality problems and safety incidents since clutter breeds problems as outlined so well by Mr. Suzaki in the earlier chapter. Let's clear the record, the motivation for most cycle time reduction programs is ultimately financially motivated. However spin-off benefits directly affect other areas such as customer satisfaction and indeed, safety.

Does cycle time reduction directly equate to the need for a safety management system? No. Does cycle time reduction contribute to improved safety? Yes, as long as the process planners are trained and build good safety practices into the new work processes and systematically replace the clutter with order and cleanliness.

Documentation System

Quality systems such as ISO 9000 require formal documentation systems. These systems are designed to translate management policy and requirements into a hierarchy of documents that provide each employee with the information needed to understand and perform his or her responsibilities completely and correctly. The normal hierarchy consists of:

- ***Policy Manual*** - *What and Why?*
- ***Procedures*** - *Where?*
- ***Work Instructions*** - *How?*

This chapter contributed by Stanley G. David, Director, Management Consulting & Rating Services, DNV-Loss Control, Houston, TX.

- *Forms & Records* - *Our organizational memory!*

The experience gained by quality practitioners and ISO 9000 Registrars working with tens of thousands of firms all over the world has taught a simple lesson regarding the need for such documentation.

> If a system has not been documented in procedures then it has not been thought through.

Too often the management of a firm thinks that hiring talented people and buying computers is enough, that these individuals will provide documents if necessary. These talented people probably do have the best of intentions. However, compressed cycle times, reduced budgets, heavy workloads, and limited access to administrative help take a heavy toll on doing what is right or needed regarding documentation. Another issue is experience and ability, even though a person may have advanced degrees or many years of management experience it does not mean that they are competent in writing useful quality procedures. First, because they may not really understand the tools and techniques of quality and secondly because they may not be good at writing concise management system documents. The negative results of people operating without procedures lead to:

- Increased cycle time, since personnel must create or recreate methods to do their jobs. If jobs change frequently and different people perform jobs then this cost can be considerable.
- Variation and errors, since without procedures personnel can only rely on memory to do work consistently. Memory does not work on complex tasks.... Would you like the pilot of the plane you are flying on attempt to land it without checklists? The

This chapter contributed by Stanley G. David, Director, Management Consulting & Rating Services, DNV-Loss Control, Houston, TX.

goal is to define the best methods to do a job and stick to them until something better comes along... then change the procedures.

- Increased training costs and time for new or transferred employees. Improperly controlled On-the-Job Training (OJT) can perpetuate errors because each succeeding group of trainees learns incorrect practices and passes them along.
- Lost organizational memory. If employee memory and personal notes are used, rather than official documentation systems, the firm looses part of their organizational knowledge each time an employee leaves or changes jobs. This invaluable knowledge is not consistently passed along in OJT. This forces new employees to relearn the expensive lessons that the previous employee had to figure out. Sometimes, depending on the seniority of the employee leaving, similar performance will never be recreated or achieved again since the employee who is leaving may have spent a lifetime developing his or her skills.

I'm sure it isn't too different with safety achievement.

Next, even when documentation has been provided personnel may not maintain or use it. Avoiding this comes down to simple management discipline. If the management of an organization takes the procedures and methods that they have paid for seriously then they themselves must show their commitment to running the farm using them. Employees will follow management's lead here. If the procedures are only "to show to auditors," then any similarity between actual employee practices and the procedures will only be coincidental. The results of this situation are very close to the ones described in the previous paragraph.

This chapter contributed by Stanley G. David, Director, Management Consulting & Rating Services, DNV-Loss Control, Houston, TX.

Common E-H-S-Q Documentation Systems

This is an area where the safety and quality management systems are totally compatible.

The above dialog can be repeated for safety but rather than nonconforming products and dissatisfied customers the results are accidents and incidents. The disciplines and technology of producing and maintaining procedures for safety are the same. In fact, some firms use common management or operating procedure systems that have safety procedures included or have the safety practices built into the operating procedures. A recent article comparing ISO 9000 and workplace safety stated:

> *Thus, if you already have an ISO 9000 based system of policies, procedures, and work instructions, you can easily incorporate new documents to cover your workplace safety program. If you are getting ready to implement an ISO 9000 system, then you can add the safety needs to the implementation process.*

The only rub tends to show up between safety and quality professionals who each try to carve out their own territory and issue their own document system. This tendency should be squashed and both moved towards company operating procedures that include disciplines from quality and safety.

Government Requirements

For many years the United States government has issued quality requirements to industries supplying it with products and services, originally starting with military equipment, then growing to all areas of federal procurement. The reigning government standard for quality has been MIL-Q-9858A and its variants. For the most part these requirements were contractual obligations that have civilian

This chapter contributed by Stanley G. David, Director, Management Consulting & Rating Services, DNV-Loss Control, Houston, TX.

parallels in automotive quality system requirements such as Ford's Q101.

The government systems required conformance with certain general system requirements that resemble ISO 9000, but the government did not elect to require any form of certification. Enforcement has been by the applicable government agency or its designated representative such as the Department of Defense DPRO activities, but enforcement was contractual and not legal. Unlike OSHA or FDA inspectors who can fine or prosecute, the government quality inspectors were limited to rejecting products or taking contractual action.

The United States Government is converting their quality system's requirements to ISO 9000 and is dropping MIL-Q-9858 and its complimentary standards. The requirement for third party certification has not been accepted as of yet since doing so would eliminate the need for government employees currently doing quality system evaluations. What will happen next is open for debate, but some form of similar compliance-based inspection approach is the most common prediction. The fly in the ointment is that the major contractors are independently pursuing third party ISO 9000 certifications hoping to bid their in-house government inspectors a fond farewell.

From the safety side, government, state, and local enforcement and monitoring has been a way of life for most companies. They work with the laws and implement their approaches to address them. This involvement has spawned the need for safety specialists and legal help to support the firm's needs in interpreting, applying, and managing safety requirements.

From the quality perspective a real need still exists to maintain separate expertise in government safety require-

This chapter contributed by Stanley G. David, Director, Management Consulting & Rating Services, DNV-Loss Control, Houston, TX.

ments within the firm. Except for the most sensitive government and regulated products the quality requirements will be ISO 9000 based and need only normal quality program personnel to meet governmental requirements. Many safety requirements are quite specific, are regulated, and require special training or expertise to meet minimum requirements. Safety related skills are not a part of current quality literature or training regimens, so most quality personnel are not prepared to deal with them. Therefore, dedicated safety expertise will be needed to ensure satisfaction of these safety specific regulations.

Inspection and Testing

Inspection and testing involve the systematic examination of the organization's purchased and manufactured equipment, components, services, materials, chemicals, and other products. The inspections and tests performed are a basic element of a quality control program and are requirements of an ISO 9000 quality system. Inspection and testing are necessary to insure that products and materials provided by the organization meet the requirements of specifications, codes, standards, and applicable regulations. Inspection and testing can take many forms, for example:

- A traditional off-line human verification using instruments and gauging
- Automated in-line or in-process checking using computer controlled test equipment
- 100% inspection using 'Poke Yoke, or error-proofing,' which automatically verifies each operation in the succeeding step through the use of fixtures and templates which will not accept incorrect previous steps in the process.

Inspection and testing, by its very nature requires a judgment as to the level of conformance of a product or process

This chapter contributed by Stanley G. David, Director, Management Consulting & Rating Services, DNV-Loss Control, Houston, TX.

to defined requirements. This requires credibility on the part of the measurement process as being capable of such measurement and an environment conducive to successful and objective measurement.

The quality disciplines that apply to these measurements are intended to help define specific and measurable characteristics of the product. The documentation created by such testing becomes the evidence of successful accomplishment of this quality level. When products fail such evaluations against standards, the resulting deficiencies become opportunities for improving the process or reducing variability and ultimately cost. Records and data generated by the inspection and testing program are a source of management and engineering information. This information is invaluable for the improvement of processes, products, quality, manufacturing efficiency, and supplier performance.

Inspection and testing requirements are usually defined in procedures and elaborated in suitable work instructions. These instructions define the responsibilities and specific criteria for and methods to be used to accomplish the inspection and testing. Procedures are normally deployed when appropriate and include record keeping and documentation requirements. Approaches to this documentation vary considerably, for example:

- On-line inspection planning and data systems take operators step by step through inspections and tests. Data recording and reporting is automatically accomplished.
- Pre-defined check sheets for operators to record a product's actual dimensions or properties.
- A quality procedures manual which identifies inspection points and refers to specific requirements

This chapter contributed by Stanley G. David, Director, Management Consulting & Rating Services, DNV-Loss Control, Houston, TX.

in an inspection handbook that defines approaches to performing specific inspection tasks.

Ideally, procedures define any environmental constraints for successfully performing inspection and testing such as temperature, humidity, contamination, and so forth. These requirements are frequently found as a part of the specific inspection and testing procedures.

Taken in context, traditional Safety Inspections are usually focused on facilities, safety equipment, safety related documentation, safety preparedness, safe handling practices, first aid and disaster planning. These inspections seldom focus on the inspection of product characteristics to determine if they meet customer requirements. However, there are everyday exceptions. In certain dangerous or critical lines of work, safety inspection is an integral part of the product, for example:

- Flight safety inspections of spacecraft by NASA to NASA 5300.4 Series S,R,QA,
- Safety certification tests by Underwriters Laboratories for electrical appliances
- Tank inspections prior to loading dangerous chemicals.

These inspections are designed and built into the normal work processes and they are performed by personnel trained to evaluate the level of safety conformance against certain guidelines, checklists, or procedures.

The similarities between a quality assurance inspector doing a quality product check and a safety inspector doing a safety inspection are many. The issue regarding merging roles is dependent on some practical issues. The most obvious are roles, training, qualifications, and economics. That is, an engine inspector on an automotive assembly line can not be expected to be pulled off line to do industrial

This chapter contributed by Stanley G. David, Director, Management Consulting & Rating Services, DNV-Loss Control, Houston, TX.

hygiene inspections in the paint facility.

Inspections, whether for quality or safety, require certain skills and knowledge that may be highly specialized. Companies must decide in whom to invest specialized training depending on their capabilities, education, desires, and potential job progression. For this reason investing many thousands of dollars of industrial hygiene training on a person who will spend their career checking engine crankshafts may not make sense. It is because of these practical issues that personnel tend to become more specialized and gravitate towards either quality or safety. As mentioned above, if the safety related inspections are closely aligned to the employee's normal areas of responsibility, then the merging of safety inspection with product quality related inspections should be considered.

The one area that the logic of combined inspection does not support is effectiveness. Inspection by personnel responsible for an area can become routine and ultimately the effectiveness of the inspections drops unless provisions are made to vary assignments and break the monotony. If an objective inspector looks at an area that he or she is not responsible for, then the results of the inspection can sometimes improve. This is due to paradigms that we build or mental rules that we establish in areas that we know a lot about. We tend to not see changes or things outside our mental field of view. Someone independent can come into the same area that a person from that area has inspected and see many issues because they have no such paradigms. This is another reason for organizations to hold independent safety inspections at some frequency. Ultimately the organization must balance efficiency, effectiveness, and cost when deciding how to, or if to, combine some level of safety and quality inspections.

This chapter contributed by Stanley G. David, Director, Management Consulting & Rating Services, DNV-Loss Control, Houston, TX.

Internal Audits

Most organizations have established policies, guidelines, instructions, and procedures designed to help manage the organization's effectiveness. The approaches for developing, implementing, and maintaining the documentation and management systems can vary greatly. Regardless of the approach *if the documentation and management procedures systems are not consistently maintained and periodically scrutinized they will degenerate.* The maintenance of the quality system is therefore very important and quality audits are seen as a key management tool in maintaining a quality system.

A quality audit is defined in ISO 8402 as "A systematic and independent examination to determine whether quality activities and related results comply with planned arrangements and whether these arrangements are implemented effectively and are suitable to achieve objectives." Since quality system audits are performed in a structured way, the preparation of audit plans based upon the requirements and scope of the standards or policies being audited are usually documented. The audits are typically scheduled and planned according to the importance of the activities and the results of previous audits.

Execution of the audits frequently include the use of checklists, reporting of findings, and planning of corrective action. The corrective action or follow-up activities are usually done in a documented and controlled manner. Records of the planning, audits, findings and corrective actions are usually maintained. Communication of results is provided to the affected areas, the quality system management representative and to management for evaluation of the effectiveness of the quality program.

Audit results are totally dependent on planning and the auditor's talents. The previous paragraphs dealt with prepa-

This chapter contributed by Stanley G. David, Director, Management Consulting & Rating Services, DNV-Loss Control, Houston, TX.

ration, however the qualifications of the auditors themselves must be considered. Auditing is a communications process, so people skilled in communicating effectively with all levels of the organization make better auditors than those who do not communicate well. Also auditing is an inquisitive process, so persons who exhibit inquisitive and deductive reasoning skills tend to be more effective as auditors.

In addition to basic qualifications, the auditors need training in quality systems auditing and the relevant company auditing procedures. The training should test the auditors' observational, communications, reporting and personal audit conduct skills. Records of this training are a requirement of quality systems such as ISO 9000.

The frequency of the quality system internal audit program required for any area depends on the situation. If an area is having difficulties implementing or maintaining its quality discipline then it is audited frequently to see that committed actions are occurring. As a rule of thumb, the whole system is usually audited every year.

The personnel performing quality system audits must be independent of the function audited. This is reasonable because it is easier for the auditor to evaluate how a system is working if they are not intimately involved or biased with the process. Independent auditors have different paradigms and can offer new approaches or recommendations. Finally, independent auditors have less political or organizational pressure on them and can operate objectively when facing negative situations.

When establishing a safety audit program, the same framework described above can be used. Remember, this is a management system audit and not inspections as described in the previous section. The auditors are looking to see if

This chapter contributed by Stanley G. David, Director, Management Consulting & Rating Services, DNV-Loss Control, Houston, TX.

the overall safety management system is in place, documented, effective, and is meeting the requirements and procedures established for it. With some minor safety training most personnel who perform quality management system audits can perform a similar service in the safety area. *Setting up a totally separate safety management system's audit would be unnecessary duplication.*

> The internal management systems audit program is well suited to support auditing of Quality and EHS systems.

Management System Certification

Currently a lot of efforts have been spent on certifying quality management systems, for example, ISO 9000. This system has the advantage of having an independent third party look at the firms from the quality systems requirement's perspective and judge them without bias. This is why the global marketplace has adopted ISO 9000 since it

ISO-9000 Certifications Worldwide

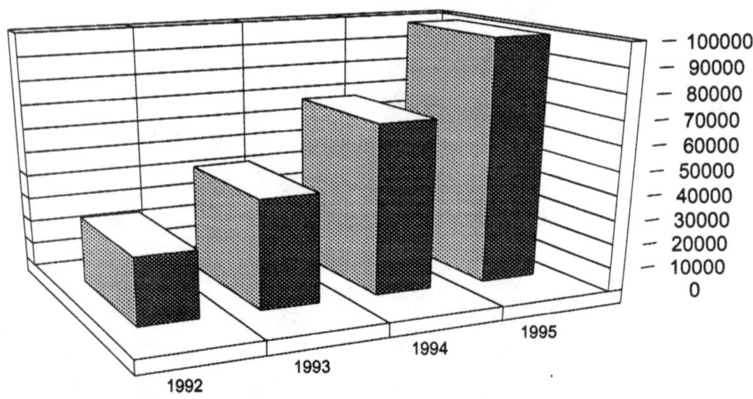

Source: Mobile Europe, Ltd., (1995 Projected)

This chapter contributed by Stanley G. David, Director, Management Consulting & Rating Services, DNV-Loss Control, Houston, TX.

is a way of reducing the risk of the customer in dealing with a supplier. The customer can be assured that the supplier has at least ISO 9000 levels of management discipline if they maintain their certification with credible registrars. Many companies have pursued ISO 9000 management system certification. The following graph shows that worldwide certifications have 100,000.

The experience of management system certification is very positive. That is, many firms who had very weak internal disciplines were challenged to improve them once exposed to third party certification. In the past they had convinced themselves that their systems were adequate but when compared objectively to the requirements of a management system standard such as ISO 9000 they were embarrassed with their performance. The advantage of such a system is that the third party certification process represents the marketplace and can sometimes provide invaluable feedback to a firm. If safety management system certification were to be accepted by the global business community, then similar results would be achieved in the area of safety. A much safer global workplace would be the result.

As of this date no marketplace equivalent has been constructed for safety. In its absence, Det Norske Veritas and the International Loss Control Institute have established the widely accepted International Safety Rating System (ISRS). Similar benefits could be gained by the use of ISRS or if the international community accepted a safety management system model and promoted it as it did ISO 9000. Again, it is rumored that such a document is being developed by the International Organization for Standardization (ISO) and will be released in 1999.

Nonconformity Elimination

Since nonconformities or deviations from requirements are

This chapter contributed by Stanley G. David, Director, Management Consulting & Rating Services, DNV-Loss Control, Houston, TX.

the primary symptom of an issue or problem, the quality management system must have mechanisms to identify and analyze these issues. Organizations usually have some form of nonconforming material reporting system that identifies and documents these issues for analysis and action. Some organizations have sophisticated data systems that clearly track and classify all anomalies from desired parameters. Others have more simple systems in place. Regardless of the elegance of the system, the organization should have an effective means to deal with nonconformities. Philip B. Crosby offers the following advice regarding nonconformities elimination:

1. *Mistakes are caused by two factors: lack of knowledge and lack of attention.*

2. *Knowledge can be measured and deficiencies corrected through tried-and-true means. Lack of attention must be corrected by the person himself or herself, through an acute reappraisal of his or her moral values. Lack of attention is an attitude problem. The person who commits himself or herself to watch each detail and carefully avoid error takes a giant step toward setting a goal of zero defects in all things.*

This means that eliminating nonconformities requires the ability to define the problem, assess severity, assess trends or recurrence, and then define responsibility. The typical quality system not only has the capability to address internal production problems, it can identify other potential sources of nonconformity such as supplier material and services, engineering generated nonconformities and tooling and process control systems. Even customer generated nonconformities such as defective customer supplied products are covered and addressed by these systems.

The nonconforming material control system itself is not the objective. The elimination of the sources of the noncon-

This chapter contributed by Stanley G. David, Director, Management Consulting & Rating Services, DNV-Loss Control, Houston, TX.

formity is. It is therefore important that those who act to correct problems are properly trained. It is a common misconception that a person doing a task for a long time is capable of solving whatever problems come his or her way. This is not necessarily the case since many problems have root causes that may extend beyond the skills or scope of an individual. An example is a process operator who attempts to compensate for variability in a process that generates a large amount of defective product. Upon investigation the source of the variability is contaminated raw materials being used up to reduce costs. The operator's part of the process many be hundreds of steps into the process and he or she may have no knowledge of the activities occurring when the raw materials are introduced into the first stages of production. This example is indicative of two areas which should be addressed by the corrective action investigation and follow-up systems:

- Training of personnel so they are able to investigate and analyze the root causes of problems, and
- The need for management involvement when problems extend beyond the scope of an employee's span of control.

Training personnel in investigation and follow-up techniques frequently includes training in applicable company procedures, use of forms, helpful graphical and statistical analytical tools, reporting procedures, corrective action assignment protocols, project management techniques, and corrective action effectiveness verification requirements.

Corrective and preventive actions are a never-ending search for ways to improve every activity in the company. In many companies it is the role of all employees, including management, to improve quality, safety, productivity, costs, profitability and business growth through the elimination of problems or nonconformity. Since customer expectations

This chapter contributed by Stanley G. David, Director, Management Consulting & Rating Services, DNV-Loss Control, Houston, TX.

continue to increase, businesses are constantly challenged to make the changes needed to satisfy the changing needs and expectations of internal and external customers.

In more effective organizations, corrective and preventive actions are focused and formalized. When it is not, experience shows that resources are often expended waste-fully. This lack of formalization, for example, records is a source of organizational amnesia. In time people are moved or forget, and bad habits creep back in bringing the same old problems back again and again.

The safety management system corollary to nonconforming material, is safety incidents. These incidents represent a deviation from desired behavior that led to an unsafe act or practice that may or may not have resulted in damage, injury or other loss. A good safety system also documents these issues and similarly reports, analyzes, and corrects the source of the incident or loss. If data systems, forms, and procedures are in place for the nonconforming material system, then it makes sense to use this infrastructure for safety related issues. Due to the format of some systems, the forms and reporting may have to be adapted for each to accomplish their missions.

> However, a management problem is still a problem no matter if it be quality or safety and it must be addressed by the appropriate authority. Therefore, if there are existing systems that collate and summarize issues such as nonconformity and safety incidents, integrated systems for management reporting and corrective action should be considered.

Process Control

Process control is a series of disciplines that the organiza-

This chapter contributed by Stanley G. David, Director, Management Consulting & Rating Services, DNV-Loss Control, Houston, TX.

tion should have in place to ensure that the manufacture and delivery of products is done at the least cost, with the best quality, in the shortest time possible. Process control involves the systematic planning of the organization's production and services that directly affect the quality of these products, to ensure that they are carried out under controlled conditions. These disciplines are extremely important in that without pre-planned and controlled conditions a great deal of variability is introduced. This variability ultimately shows up as increased costs, reduced quality, and poor delivery performance.

Process control planning is a simple concept. Essentially, production process planning is simply the sequence of production or test operations called for on the process routings or work instructions. Since most organizations have some sort of process instructions or manufacturing planning documents, process planning requirements can simply be these process routings plus related work instructions. The generation of these documents is a part of process as well. An example is a machinist who is required by a simple manufacturing routing to machine, check and record specific dimensions on the parts he or she is running, using the route card. In more complex production or process environments, process control planning can be formalized documents that carry the responsible personnel through a sophisticated series of steps and record requirements. An example here is assembly technicians building a diesel locomotive against fully detailed manufacturing drawings and written assembly process requirements.

Whatever the method or format for control of processes, most manufacturing and complex service firms plan out and execute these activities as a part of the product life-cycle. From raw material inspection, through manufacture, then on to functional testing at a customer's installation, a pro-

This chapter contributed by Stanley G. David, Director, Management Consulting & Rating Services, DNV-Loss Control, Houston, TX.

gressive series of pre-planned operations and evaluations is designed and sequenced to provide confidence that the products meet all applicable requirements. Process control is a technical value-adding step in the production process and provides for adequate resources, computerization, equipment, instructions, records and qualified personnel. If an operation is properly planned; correct materials provided; all equipment is working properly, calibrated, and capable; the environmental conditions are appropriate; and the employee performing the task is trained, qualified, and competent; then there is a good probability that the task will be performed successfully. If one or more of these preparations are not fully addressed then there is an equal probability that errors will occur. There is a large body of work in the quality field that address these disciplines that add together to achieve process control.

Inspection and test plans may be a part of the production process control planning or be specific unto themselves. However, if not a part of the production process planning, then related or referenced work instructions are usually available to guide the responsible person through the required procedure. An example is a process plant that calls out the need for a test of each batch at a certain stage in the process. The process planning is a flow chart and the test is identified only by name. When the sample is provided to the lab, the technician has a documented, specific lab procedure that outlines how the test is to be performed. Additionally these plans often classify the type of test to be run and its frequency. Finally, these work instructions, planning, or related documentation such as the quality procedures, frequently identify the proper steps to be taken if products fail these required inspections or tests.

This chapter contributed by Stanley G. David, Director, Management Consulting & Rating Services, DNV-Loss Control, Houston, TX.

> From the process control perspective, safety management is very much interwoven into its basic disciplines. In many organizations, when a new product or process is being planned, safety considerations are considered as carefully as are quality or technical requirements.

When a manufacturing or process engineer is looking at the sequence of steps needed to successfully make a product, good planning practice requires the engineer to look at ergonomics, weights, handling, and all other forms of hazards that can be created from the process. Most of the requirements and steps that are generated are not looked at as a quality step or a safety step, they simply are necessary steps. This is an area where both quality and safety are very compatible and coexist without prejudice.

Good manufacturing practices are synonymous with process control and are fundamental to good quality and safety performance.

Training and Employee Qualification

Modern total quality training strategy encourages the organization to develop and realize the full potential of all employees by maintaining an environment conducive to full participation and personal growth. Training, empowerment, teamwork, employee involvement, motivation of the workforce and performance measurement all are very important for the ongoing success of an organization. The activities needed to accomplish these objectives must be planned, defined, researched and supported by management. TQM (Total Quality Management) experience has demonstrated that a robust human resource program in the area of training pays many lasting benefits beyond the simple qualification of employees to perform assigned tasks.

One of the fundamental requirements in a quality system

This chapter contributed by Stanley G. David, Director, Management Consulting & Rating Services, DNV-Loss Control, Houston, TX.

such as ISO 9001 is the need for training personnel who directly affect the quality of products and services delivered. This training is needed for operations, management, and support personnel and covers all aspects of the work being performed. Whether controlled centrally or by each department, an assessment of the training requirements for each job is expected to be available. Training assessment is considered to be an ongoing process that continues to test the training needs of employees on a regular basis. Recurring training is typically implemented wherever it is appropriate. Training plans, course materials, training records and procedures covering training responsibilities are all components of a robust system.

Various formats and styles of training may be in effect for the organization, for example:

- A corporate training program trains all employees in the principles of total quality management
- A plant program teaches employees problem solving tools
- A department trains its personnel in corrective action procedures
- A team trains its members on proper cleanup techniques
- The quality system training objective is for the overall training process to build and maintain the skills that are needed by associates for the future as well as today.

Qualifying personnel for positions is critical to job performance, productivity, safety, quality and employee morale. In order that qualified personnel can be put into a position, some preplanning, training and verification must occur. First, the job must be analyzed and some written description prepared for the tasks to be performed. Also, the

This chapter contributed by Stanley G. David, Director, Management Consulting & Rating Services, DNV-Loss Control, Houston, TX.

technical, educational, physical, and experience requirements must be defined. Procedures for the construction and maintenance of job descriptions are considered especially important.

Firms select whomever they deem appropriate for a position. However, personnel selected may not meet or exceed the criteria defined in the job descriptions. If they do not, then what specific training and support is provided to make up for the experience deficiencies? Has the firm established qualification criteria or an evaluation period where the employee's performance in the job can be assessed? Is there a formal skill certification program required for certain tasks that require it, and has the program been planned, conducted and maintained consistent with applicable codes? Is a similar process used for the qualification of professional personnel, such as design engineers or scientists? All of these areas, if appropriate, should be a part of the of the employee qualifications process. Once initial qualifications criteria are met, quality system philosophy directs that the ongoing training and employee development program should be integrated into a long-term process that maximizes each employee's contributions and personal growth.

When comparing the safety management view of training with quality we once again find a management system that both quality and safety agree upon. Safety management expects that job responsibilities are defined, training needs identified, and an ongoing program to maintain or improve the skills and qualifications of personnel be established. There are virtually no areas relating to this general subject of training and employee qualifications from a quality systems perspective that are not transferable to the safety management viewpoint. Both are focused upon reducing the potential for loss or other problems by the proper

This chapter contributed by Stanley G. David, Director, Management Consulting & Rating Services, DNV-Loss Control, Houston, TX.

assignment, qualification, training, certification, and development of employees.

THE BENEFITS AND PITFALLS OF QUALITY

Now that we have explored the background and dimensions of quality, it is worthwhile to summarize some of the experiences gained by organizations which have utilized quality management systems. First, any quality management systems activity represents an investment by management to incorporate a selected set of disciplines into organizational practices. Whether mandated by the customer or voluntary on the part of management, decisions must be made. Even the lack of a decision is itself a decision. Some management teams have gotten and accepted the quality message while others have not. Some organizations badly need formal quality disciplines in order to remain viable. Others have such high levels of discipline already built into the management of the operation that there is very little that a formal quality program could do to help the firm. Regardless of the reasons why the portfolio of quality services for a firm was selected, the critical factors from a management perspective are:

1. Does the quality program selected support the overall strategy and financial goals of the organization
2. Can the firm afford this total program now or will it need to be phased in place over a period of time
3. Will the management team use and fully meet the demands of the quality program
4. Can we get the buy-in and support of the entire organization?

If these questions are not asked and answered affirmatively

This chapter contributed by Stanley G. David, Director, Management Consulting & Rating Services, DNV-Loss Control, Houston, TX.

then there is a good chance that all or some of the quality program will fail in time (Quite similar again to success in safety). It is for this very reason that the statistics quoted by the popular press on failed quality programs are as high as they are. The most critical of these areas is buy-in and support by the people who do the work. A cautionary note in this regard comes from management guru Stephen R. Covey:

> *So many organizations tell us that the real problems of quality have to do with the inability of people - individuals, departments, plants, entire organizations - to work effectively together. There is a root common cause to this problem in all of its manifestations...lack of synergy, failure to listen or communicate effectively, internal win-lose competition and turf defensiveness, unclear or dissimilar strategic priorities, uncertain mission and fuzzy vision - all of these and similar problems reflect a single characteristic within each individual underlying every other failure...it is the failure to "be proactive".*

When done correctly, quality programs or quality management systems can be invaluable to a firm. Oftentimes the process of implementing a quality system forces the firm to look at its management practices and disciplines and correct a lot of weak areas.

> Knowing how true this is of safety management, it's not difficult to understand why Alcoa's vice president for quality made the comment in a 1990 issue of *Forum* that he can no longer tell where quality leaves off and safety begins.

One of the most common phrases that the author has heard from executives after they have achieved ISO 9000 Quality System Certification is, "We should have done this a long time ago, the process really helped us get our act together."

This chapter contributed by Stanley G. David, Director, Management Consulting & Rating Services, DNV-Loss Control, Houston, TX.

Too often a firm gets so involved in trying to satisfy customers and stockholders that they ignore their infrastructure for too long and it gradually gets less effective. As this problem develops, the firm's personnel attempt to address more and more problems. With inefficient and undisciplined systems, margins will erode until someone orders the organization to take a time out to update the management system. Since most quality management systems form a basic set of disciplines, the establishment of such a system is very important for growing firms. The system then allows them to manage larger volumes of activities and issues without adversely affecting performance.

Finally, there is a huge number of quality tools and methods available for organizations to select from and use. Like any toolbox in the right hands these tools can do wonders, while in unsure hands they can be wasteful and actually do damage.

Some managers underestimate the sophistication of some quality tools or processes and overestimate the capabilities of their employees. The author has faced technical customer inspection and testing requirements that were frequently beyond the state of the art. In more than one case, scientists in major research universities had to be employed to come up with the necessary technologies.

For the tools and systems of quality to be used properly, careful consideration should be made regarding employee qualifications, training, experience and temperament. There is a classic story of the $40M chemical firm who ordered a line inspector with no formal quality, management, or systems experience to implement ISO 9000 in the firm. The results were the resignation of the inspector, out of total frustration, and a fragmented, ineffective quality system that was accepted by neither the organization nor

This chapter contributed by Stanley G. David, Director, Management Consulting & Rating Services, DNV-Loss Control, Houston, TX.

the ISO 9000 registration body.

On a more upbeat note, these tools can be very effective. Motorola Incorporated has retaken a large share of the world's electronics and communications equipment market as a result of an organization-wide commitment to their total quality program. It is based upon employee education (Motorola University) and on the reduction of errors (6 Sigma). Scores of firms have made tremendous leaps in performance using statistics and design of experiments, while many more have rebuilt their companies and their markets using the teachings of Deming, Juran, Shingo, Hammer, and others. Used appropriately, the tools of the quality trade can be some of the best contributors available to a firm to reduce costs, cycle time, overhead, and customer complaints while improving customer satisfaction... short of redesigning the firm and its products. As stated in a recent article:

> *Investors from Wall Street to Main Street know a good thing when they see it: quality pays. In fact, the payoff can be impressive. Stock shares of the Malcolm Baldrige National Quality Award winners or their parent companies have, overall, registered hefty returns since I began tracking them in 1991.*

THE INTEGRATION OF QUALITY AND SAFETY

Due to the growing trend towards re-engineering and general reduction in support and indirect expenses, companies have been scrutinizing quality and safety programs , among others. It is in the area of reducing costs that firms have started to look at the relationships between safety and quality. The logical question becomes, "Why support two different efforts when they can be combined into one?"

This chapter contributed by Stanley G. David, Director, Management Consulting & Rating Services, DNV-Loss Control, Houston, TX.

The relationship itself has three distinct dimensions that the management team must consider and address before attempting a merger:

- *Political* - Both quality and safety programs are laced with a high degree of emotion and philosophy that tend to create zealots. These zealots can range anywhere from the Chairman of the Board, to the quality or safety manager, to line employees, to unions, to customers, and even to the local community. Certain elements of these programs can become 'sacred cows' and any attempt toward radical change would be viewed negatively. One of these elements is independence and the voice that it provides. Also there is the issue of turf. Oftentimes the quality and safety organizations have established their own turf and have formed alliances. Any attempt to merge such entities will revert to some form of turf warfare.
- *Philosophical* - Quality programs and management systems are structured to attack quality issues. The same can be said for safety programs and systems. The popular literature, courses, and professional organizations that support the technologies for these two disciplines focus on the perpetuation of their own sciences, not their cross breeding or integration. Therefore, a lot of followers who have bought into these messages have strong opinions regarding their safety or quality missions, and will tend to dismiss the notion of blending with another discipline. Philip B. Crosby states in his book, *Let's Talk Quality*, "Safety is a great analogy for understanding quality... Everything safety is about relates to the absolutes of quality."
- *Technological* - There are certain safety skills and

This chapter contributed by Stanley G. David, Director, Management Consulting & Rating Services, DNV-Loss Control, Houston, TX.

knowledge that are indigenous to safety and must be developed and invested to perform their intended purpose. The same is true for many elements of quality technology. Competency in these specialized areas may take a significant investment of time and money or may need to be focused upon certain specialists. Besides the personnel issue, equipment must be considered as well, since specialized equipment may require special skills, for example, a wet chemistry quality assurance lab or a Halon fire suppression system in a computer room.

Firms that learn to address these issues can move toward the goal of integrating quality and safety. To do so properly takes careful consideration, a new organizational design, training, good communications, and a leader for the combined entity who can meld and balance the needs of both disciplines. In a way, the task is like forming a combined dental and medical office. Very strong attitudes, and in some cases, prejudices exist that can make the logical sharing of infrastructure a nightmare to manage. The careful selection and support of personnel with compatible personalities is an absolute necessity.

From the infrastructure perspective, many similarities exist that can be shared or merged. For example, the quality management system requires many specific controls and disciplines that are the same for safety, albeit by different names:

- Policy, procedures, and work instructions
- Process control and process planning (Including work design)
- Internal Audit Program
- Corrective Action Program
- Employee Qualification and Training

This chapter contributed by Stanley G. David, Director, Management Consulting & Rating Services, DNV-Loss Control, Houston, TX.

- Management Involvement
- Handling, Storage, Packaging, and Delivery

Other disciplines are unique to either safety or quality. These areas of difference would require their own infrastructure. Many of these were discussed above in the quality dimensions, so they will not be repeated except to say that they do exist and must be supported separately. This does not discredit the goal of integration. It sets some constraints on it. Rather than being a safety program merged into quality or vice versa, the results of integration would more likely have the following characteristics:

- *Management System* - Rather than have several departmentally based systems operating in a loose confederation, the structure would more likely become a single management system with integrated procedures, meetings, corrective action, and audit programs.
- *Pooled Resources* - The standard inspection, audit, training, and safety inspector roles would overlap. Cross-trained resources could perform both safety and quality related activities depending on the need. This would reduce some duplication.
- *Specialized Resources* - Some specialized skills and tasks would remain, such as Quality Engineering, Safety Engineering, Laboratory and Test Technicians, Industrial Hygienist, and other highly technical roles that are not appropriate for less skilled employees. These specialties would need to be staffed and supported.
- *Visible Leadership of Both Dimensions* - In combining safety and quality, the firm must not lose focus on either dimension. The need for efficiency must not be at the expense of effectiveness, otherwise the

This chapter contributed by Stanley G. David, Director, Management Consulting & Rating Services, DNV-Loss Control, Houston, TX.

> overall negative impact can be very costly. People watch management closely and if it looks as if one or both programs are falling into obscurity then the line employees will reduce their focus on them. In some ways, management will need to work harder in the combined program since they will have to make sure that the total quality and safety message is communicated and accepted. Training, reinforcement, and leadership by example must be provided in ample doses to get the total message across and have it understood and supported.

There is little evidence to support that the integrated system will be more effective from the quality or safety perspective. However, if parity in the disciplines is maintained, then some economic benefits may be gained. This benefit will include a more agile and highly utilized set of support functions with a need for less total resources.

ECONOMIC IMPACT

The economics of quality are fairly well understood. As long as we do not exceed the level of value that a customer is willing to pay for, we can improve our profitability through gaining marketshare via offering improved goods and services at market prices with reduced internal net costs. The quality dimension focuses on improvement of both sides of the accounting ledger. That is, by improving product quality it allows marketing to get or maintain higher prices, and by improving processes it helps reduce the cost of goods sold. Many of the tools that were discussed in earlier sections described how these quality processes work and what their areas of focus were. The current trend is not to focus on the tools as a means unto themselves, but to apply them appropriately in a quality based cost management system.

One of the key areas of quality based cost management is

This chapter contributed by Stanley G. David, Director, Management Consulting & Rating Services, DNV-Loss Control, Houston, TX.

cost-of-poor-quality (COPQ). This metric is a means for the business to focus improvement efforts on increased customer satisfaction and profitability. The entire organization must learn how the COPQ affects it, through the use of a cause-and-effect process. The cost-of-poor-quality is the sum of the costs associated with internally and externally generated waste, plus the cost of the infrastructure and materials needed to manage them. The methods for computing COPQ are the subject of many books and training courses. However it is important to say that, however the firm computes this cost, it historically is underestimated. A recent book *Linking Quality to Profits*"gave case after case where prestigious corporations had underestimated the COPQ by orders of magnitude. Once these firms identified these costs they began to systematically attack them and improve their profitability.

Making the changes needed to reduce COPQ does not always come easy. Many firms live with reduced efficiency and profitability rather than take on the challenges that significant quality improvement can require. Frequently the true start of large scale quality improvements is triggered by increasing competition in the market place. Sometimes this pressure forces fundamental changes in the business such as changes in management, consolidation of production, reductions in overhead, and increased process effectiveness.

The tools of quality are often employed to help transition the business into this cheaper, faster, better environment. When such shifts occur massive changes can happen in short periods of time. An example is the shift from vertically integrated firms that make virtually everything that is needed to build a product, to outsourcing where the firm only does some selected assembly and finishing operations. The rest left up to a chain of suppliers. Ultimately the

This chapter contributed by Stanley G. David, Director, Management Consulting & Rating Services, DNV-Loss Control, Houston, TX.

change becomes a cultural change and the new and improved business operates and looks significantly different compared with where it was.

Looking at safety from the quality perspective, safety can contribute to the improvement process if it is viewed from the loss control standpoint. The view must not be "lets do what is needed to satisfy OSHA." It must become "lets do what is needed to reduce the overall risk of safety related incidents and the resulting losses to people - property - environment generated by these incidents." The first view is the typical view of a quality and operations professional, since most are not schooled in the philosophies of Safety and Loss Control Management (LCM).

> Those who do understand the teachings of LCM realize that Safety has a very important role in reducing the costs of operating a business. As emphasized throughout this book, accidental loss can have a significant effect on the bottom line.

It is also true that both systems sometime claim the same gains. For example, in order to plan out a new production process the Process Engineer must identify the sequence of work. Consequently he identifies the tools to be used, any necessary training, inspections to be made, sets up procedures, identifies safety items, and in the process cleans up both quality and safety issues. Ask the quality engineer and he or she will tell you that process control, operator inspection and testing, set-up approval, work instructions, and training are all part of the quality domain. Ask the safety engineer and he or she will tell you that the job design, training, safety checklist, work instructions and controlled conditions are part of the safety domain. Ask the process engineer responsible for the process and he or she would probably say that it is normal process planning pro-

This chapter contributed by Stanley G. David, Director, Management Consulting & Rating Services, DNV-Loss Control, Houston, TX.

cedures in which inputs from various areas are sought out to ensure the process will work and satisfy everyone. So who is correct?

It is the process engineer who is probably most correct. Safety and quality professionals may contribute or significantly enrich the process but it is the company that should claim the benefits, not quality or safety. If both of these areas do their job correctly no one will get hurt, no defective products will be made and the process will work as planned, the first time and every time. The authors have seen all three areas claim savings for the same process elements as they strive to show their value to the company. A lot of gamesmanship can occur when talking about improvement numbers. Management and accounting must be wary of improvement figures and use reliable accounting disciplines to identify the sources of cost reduction and cost avoidance.

INDUSTRY REACTIONS

Quality is an integral part of the infrastructure of most companies. Quality can take the form of a formal program with full time quality professionals or an informal process with employees using selected quality tools as a part of their day to day efforts. Many studies and polls have been taken to measure the extent of use of quality tools and management processes. The data tend to indicate the following:

- A large percentage of companies in the world have adopted quality management systems, i.e., ISO 9000, QS 9000, and so forth.
- Most larger corporations have mature and long standing quality systems that have grown beyond quality assurance and exhibit many total quality management characteristics.

This chapter contributed by Stanley G. David, Director, Management Consulting & Rating Services, DNV-Loss Control, Houston, TX.

- Some industries have adopted minimum quality standards as the industry norms, i.e., the automobile industry with QS 9000, and the chemical industry with ISO 9002.
- Many smaller second and third tier suppliers to larger firms are pursuing basic quality management systems via ISO 9000 certification.
- The firms who have focused on quality as part of their basic business strategy have matured their quality systems the most and have shown the greatest returns on their quality investments.
- The firms that have adopted a 'minimalist' approach to quality have received little economic benefits from investing in quality.
- The large number of potential quality tools tends to confuse users. Best results occur when a small number of tools are well understood and are used frequently by the organization.
- Firms that become quality benchmarks such as Malcolm Baldrige National Quality Award winners tend to have superior stock price performance as compared to other firms of similar size in the same industries.
- Most firms still have huge amounts of waste that can be identified and addressed using quality methodologies.

The overall reaction is therefore one of continuing support for the use of quality as a business process and the general impression is cautiously supportive. Firms are taking their quality programs more seriously and are not trying to take on mass quality training campaigns. The current reality is to go slowly and cautiously with things that make a difference to the customer and the bottom line rather than

This chapter contributed by Stanley G. David, Director, Management Consulting & Rating Services, DNV-Loss Control, Houston, TX.

buying into a wholesale total quality management philosophy. Much of the "hype" of quality that surfaced in the last decade appears to be replaced with a more rational approach. The tools of quality are being used more, but people are talking about it less, since the tools are becoming part of normal business processes rather than "silver bullets" for special cases.

Much of the improvement that is taking place in the field of quality is due to the cultural change that has been spawned by the global ISO 9000 movement. Many firms have finally accepted the reality that the global marketplace will not accept inefficient, non-responsive, and marginal quality producers for very long. These changes have forced change on the affected firms. Some of the change that has occurred has been too much for some people in the firms and has resulted in many changes of faces in key positions. This is the reality of the cultural changes that result from the insertion of a quality management system in a largely undisciplined firm. Some employees cannot or will not adapt, and management is forced to make hard decisions. The net effect has usually been positive in that more adaptive and forward thinking personnel have succeeded the inflexible thinkers. The firm is usually stronger in the long run. For the most part, industry is meeting these new challenges and is succeeding in the painful transition to this new level of performance. Firms now have the opportunity to justify safety improvements in conjunction with the changes that are being made to the quality systems.

> With the tightening of OSHA rules and movements in the international marketplace to adopt global safety standards, firms are taking another look at their organized safety programs. Some are starting to look at the benefits of integrating safety and quality management and a few have

This chapter contributed by Stanley G. David, Director, Management Consulting & Rating Services, DNV-Loss Control, Houston, TX.

actually taken steps to implement such systems. It is the author's view that this process is still in its early stages. Much is still to be learned about how to do this integration effectively. One of our key goals should be to learn as much as possible from those who attempt this implementation in order to learn whether this integration is for us, or how to accomplish it most effectively.

CONCLUSION

From the quality perspective, quality tools and management systems are alive and well on a global basis. The growth of quality management system certification such as ISO 9000 has forced thousands of firms who were marginally committed to quality into a more proactive stance. The effect of this global quality movement has been to institutionalize basic quality disciplines and force a net improvement of effectiveness, efficiency and quality on the part of suppliers of products and services. This has allowed more players who were once marginal to compete and to challenge the previous industry quality leaders.

The net effect of this global quality normalization is more competition, better prices, and larger markets for those who survive this competition. Therefore, from the consumer viewpoint, better products, more variety, better service and more competitive prices are the results that should be expected. From the quality practitioner's viewpoint, it means a huge jump in the demand for support of quality expertise and service to support the technical needs of these firms entering the global quality marketplace.

Safety tools and disciplines complement quality for the most part and can utilize a common infrastructure if planned properly. Procedures, training, planning and so on do not care if the subject is quality or safety. The systems

This chapter contributed by Stanley G. David, Director, Management Consulting & Rating Services, DNV-Loss Control, Houston, TX.

are compatible. In certain technical areas there is a strong argument for maintaining separate expertise. Along with the technical idiosyncrasies, the political and philosophical differences between the programs as defined by the firm must be considered. These areas must be identified before attempts are made to do any integration. Like quality, safety methods and tools have their own set of issues when considered as components of a quality-safety management system. For example:

> *There's a definite distinction to be made between workplace safety and product safety. ISO 9000 incorporates both to a certain extent, but more attention is paid to product safety, which is clearly specified under both design requirements and document control. You will have to pay careful attention to the language of the standards in deciding where safety issues should be incorporated.*

Safety is growing in visibility and has been pushed into a more prominent role by regulations such as OSHA. The net effect of OSHA is similar to that of the Department of Defense Military Standards, firms grudgingly complying with the requirements but not embracing them as a tool to drive success. For this reason, the OSHA rules are doing a bit of disservice to the industrial safety marketplace. If the International Office for Standardization in Geneva does as good a job in crafting a safety standard as has been done with the ISO 9000, then the safety management evolution will switch from government conformance to marketplace disciplines. In looking at the ISO 9000 model the marketplace can be much more demanding and ruthless than any government agency will ever dare to be. Then safety will be strengthened, as quality has been by ISO 9000.

This chapter contributed by Stanley G. David, Director, Management Consulting & Rating Services, DNV-Loss Control, Houston, TX.

Selected References

Berkowitz, Ralph. *The Third Generation of Quality*. The Quality Management Forum. Summer, 1995.

Covey, Stephen R. *The Demand to Be Proactive*. Covey Leadership Center, 1995.

Crosby, Philip B. *Quality Without Tears*. McGraw Hill Book Company, 1990.

Environmental Management Systems. ISO 14000. ISO/TC 207 Committee Draft, February 1995.

Goldratt, Eliyahu M. *Theory of Constraints*. North River Press, Inc., 1990.

Helton, Ray. *The Baldie Play*. Quality Progress. February 1995.

International Safety Rating System TM (ISRS). Det Norske Veritas, 1994.

Kozak, Bob. *Can Safety Issues Be Integrated Into an ISO 9000 Program?* Quality Systems Update. August 1995.

Kozak, Bob and Rick Clements. *ISO 9000 and Workplace Safety*. Quality Digest, May 1995.

Linking Quality to Profits. ASQC Quality Press. 1994.

Malcolm Baldrige National Quality Award. 1995 Award Criteria. United States Department of Commerce.

Model for Quality Assurance in Design, Development, Production, Installation and Servicing. ISO 9000. ISO, 1994.

Quality Management and Quality Assurance Vocabulary. ISO 8402. ISO, 1994.

This chapter contributed by Stanley G. David, Director, Management Consulting & Rating Services, DNV-Loss Control, Houston, TX.

Chapter 5

SAFETY PAYS

"Only a profitable business can generate or raise the capital that will enable it to grow, produce more goods, hire more people, and pay improving wages."

Frederick R. Kappel, Former President, American Telephone and Telegraph

---

That is a question that many corporate health and safety professionals wish could be discussed, or even imagined, at the boardroom level. This is not to suggest that executives do not care, or that they do not discuss safety; it is just that, in their minds, safety and health considerations frequently register as a net cost and not a net profit to the enterprise. Therefore, other business matters take higher priority.

Certainly employee safety and health is a recognized imperative from an ethical, moral standpoint in any well-managed organization, but this is seldom connected to its potential for increased profitability.

The IAPA's field staff experience, in directly speaking to

This chapter contributed by Don J. Pedley, Engineering Director, Loss Management Services, Industrial Accident Prevention Association, Ontario, Canada.

more than 30,000 organizations each year is that, as motivators, the cost of workers' compensation premiums and the jeopardy of corporate fines in Ontario, of up to $500,000 per violation, far outweigh in management's minds the fact that profits only exist to the extent that all costs, both current and potential, are fully absorbed.

Much of our difficulty in raising management's consciousness of hidden potential profits, deriving from reduced losses, may be traced to a sometimes perceived inevitability of accidents and injuries. This, in turn, possibly relates to personal experience. When we personally experience an accident, whether locking our keys in the car or cutting ourselves with a kitchen knife, as soon as we recognize the consequence, we admit it was our own fault, whether by distraction or by lack of specific forethought or care—we instinctively know we were to blame and look to our action rather than what led to it. Why then would it not also be automatically assumed to be the fault of the individual in a workplace when accidents and injuries occur? As a result, it is not surprising that this thought is carried into the executive office: not at all surprising when reinforced by the fact that over sixty years ago "the father of safety", H. W. Heinrich, reported from insurance files that 88% of all accidents were caused by unsafe acts, and concluded that these acts were brought about by "inherited or acquired faults of persons."

With these thoughts in our common psyche, then, it is quite reasonable that without some clarion call or some solid evidence to the contrary, accidents and resulting costs may be seen as inevitable and unpreventable by many decision makers. The natural result of such thinking may be to measure the cost of safety programs and equipment only against potential reductions of compensation insurance premiums or to treat safety costs as an additional insurance

This chapter contributed by Don J. Pedley, Engineering Director, Loss Management Services, Industrial Accident Prevention Association, Ontario, Canada.

against prosecution and penalties.

In the past decade when margins shrank and markets were eroded by competition, the prime culprit was seen as quality deficiencies which resulted in higher costs and disenchanted customers. This was the particular wake-up call that resulted in Quality Circles, TQM, Quality is Job 1, and the rush to comply with the new International Standards, the ISO 9000 series.

The mass movement to embrace quality management principles brought envy, combined with some hope, to every observant health and safety professional. When reading the works of the quality teachers, including W. Edwards Deming, Joseph M. Juran and Phillip Crosby, it was immediately obvious to safety people that the advice and formulae for quality management improvements could equally be applied to safety management. If quality is free, cannot truly safe and healthy workplaces contribute to profits?

The whole concept of doing it right the first time, resolving the issues upstream in the system and measuring control factor deviations from standards, rather than waiting for final inspections of consequence, so closely replicates modern safety management theory that it is interesting to ponder which came first. Of course, the answer is that neither came first—they are the same. Each is the product of basic problem-solving theory that has been taught to industrial engineers and managers since the dawn of the machine age: define the problem (improvable solution), evaluate possible solutions, choose one as the plan, institute the plan, and monitor performance.

The key step is the first: thorough definition of the problem. To identify not only the most obvious causes but all the underlying, basic or root causes. In safety, just as in quality, basic causes will invariably identify management

---

This chapter contributed by Don J. Pedley, Engineering Director, Loss Management Services, Industrial Accident Prevention Association, Ontario, Canada.

system failures (oversights) which have resulted in lack of defined standards or lack of enforcement of those standards.

In matters of quality deficiencies, measures to identify and resolve basic causes were quickly embraced in the 1980's by the major North American corporations whose markets were shrinking. Now the ISO 9000 registration banners are seen flying everywhere, evidence that quality management systems are not only in place but have been audited to ensure that they are operational and not just paper programs. With the continuing, but frequently ignored, erosion of business margins, due to the cost of injuries to people and damage to property, materials and the environment, it is clear that another awakening of management awareness on a wide scale is due, if not overdue.

It is possible, of course, that establishing audited quality management systems will automatically bring improved safety performance as an added bonus. It is difficult to envision a truly quality workplace that ignores safe work procedures or a culture which meets ISO standards but still tolerates injured employees. Apart from the possibility of safety improvements by quality osmosis, it is still desirable to make a direct case for improved corporate profitability resulting from reduced workplace losses.

Because of the number of variables in both accident causation and outcome, the measuring of consequential accident costs will probably always be empirical in nature and generalizations as to cost ratios will always abound. From a business point of view, this increases the difficulty in selling the cost effectiveness of any investment in occupational health and safety. What cannot be ignored, however, is the high proportion of attendant costs when compared to the direct or insured cost, so often the only measured criterion.

This chapter contributed by Don J. Pedley, Engineering Director, Loss Management Services, Industrial Accident Prevention Association, Ontario, Canada.

Based on a number of case studies in 1926, H. W. Heinrich, declared that "compensation and medical payments constitute only one fifth of the total employer accident cost."

In 1944, R. B. Blake, Senior Safety Engineer, Division of Labor Statistics, US Department of Labor, wrote, *"the main driving force behind the industrial safety movement is the fact that accidents are expensive. Substantial savings can be had by preventing them."*

Numerous cost models and costing strategies have been developed over the years, including the Simonds model which was published in 1955 at the request of one of the President's Conferences on Occupational Safety. Simonds' method was later recommended by the National Safety Council. A more recent model was presented in 1991 by Professor Maurice Oxenburgh of Sydney, Australia, in his book, *Increasing Productivity and Profit through Health and Safety*, with credit for original research given to Dr. P. Luikkonen of Stockholm University.

Heinrich's discussion of the ratio of insured to non-insured costs was developed in the iceberg theory introduced by Frank Bird, Jr., in the 1975 publication, *Management Guide to Loss Control.* This iceberg showed uninsured property damage as between 5 to 50 times the insured costs of injuries, with uninsured miscellaneous costs adding a further 1 to 3 times the costs of compensation and medical expenses.

In 1993, Great Britain's Health & Safety Executive published a series of five case studies to illustrate specifics of Bird's iceberg relationship of insured to non-insured costs for specific accidents occurring in various business sectors.

The cost ratios identified were as follows:

This chapter contributed by Don J. Pedley, Engineering Director, Loss Management Services, Industrial Accident Prevention Association, Ontario, Canada.

	Insured Costs	Uninsured Costs
Construction site	1	11
Creamery	1	36
Transport Company	1	8
Oil Platform	1	11
Hospital	1	8 - 36

Although it was recognized that such limited studies could not establish that the organizations participating in the research were representative of their industry sectors, it is interesting to note that all fall within the parameters of Bird's model, and all exceed the 4:1 ratio proposed by Heinrich. This supports the thought that the profits to be gained by overall Loss Control Management are large enough to merit serious business consideration.

In the Province of Ontario, Canada, the Industrial Accident Prevention Association (IAPA) is a membership organization serving over 120,000 workplaces in the manufacturing, retail/wholesale, tourism, hospitality and administrative office industries. More than 70 field consultants call daily on member firms in need of help and together have established an excellent track record of reduced injury frequencies in firms consulted with. A major incentive for the decision-makers in agreeing to improve their safety management in a systematic way, is the potential of reducing the cost of compensation premiums and penalties. For added motivation, the Bird Iceberg has for years been utilized by consultants to illustrate further potentials available for loss reductions and hence improved bottom-line profits.

This factoring up of compensation costs to illustrate potential savings, and hence improved bottom line results, has proved a telling argument in a Province where the insurance (benefits) payments to injured workers, including pensions

This chapter contributed by Don J. Pedley, Engineering Director, Loss Management Services, Industrial Accident Prevention Association, Ontario, Canada.

and death benefits, amounted to $2.865 billion (Canadian) in 1993.

Using the "Bird Iceberg" ratios, this is readily extrapolated to between $17 billion and $152 billion in overall accident related costs totally carried by Ontario industries. This is obviously an enormous burden in the fierce competitive environment of NAFTA and world trade.

1n 1993, the reported net revenue for the whole of Canadian industry amounted to $16,373 billion. Considering only the most conservative of estimates, if only 25% of the Ontario losses could be eliminated by positive preventive actions, it would clearly represent an enormous boost to competitiveness and market shares for business in the Province.

In 1991, recognizing the persuasive power of connecting reduced injuries through improved health and safety management to the reduction of losses and resulting bottom line improvement, the IAPA initiated a research project to identify total accident costs. Working with industry representatives from many sectors, Dr. James Hansen, an IAPA staff research specialist, developed a model which identifies in a simple form the insured and non-insured components of the iceberg. It is hoped in the future, with the help of member firms and national and international contributors, more data on accident costs based on the model will be collected. Once data are available in sufficient numbers, it should be possible to group by sector and by firm size to provide a most persuasive view of hidden costs and hence potential profits.

For the costing model, Hansen has classified costs associated with accidents in two major categories: *objective* and *subjective*. He notes that *objective* costs are those which are able to be calculated, some more readily than others. *Sub-*

This chapter contributed by Don J. Pedley, Engineering Director, Loss Management Services, Industrial Accident Prevention Association, Ontario, Canada.

jective costs, however, are not able to be calculated in financial terms but should be recognized. *Subjective* costs include such things as loss of company public image as a safe place to work, reduced employee morale, and psychological impact on supervision.

The costing model classifies *objective* costs under five headings: compensation and benefits costs; legal costs; time/productivity and production costs; material, equipment and property damage costs; and miscellaneous other costs. Since the research is ongoing, the list of cost items may not be complete. But to highlight for decision makers the comprehensive nature of potential losses due to accidents, and the resulting potential savings available through an effective loss control management program, Dr. Hansen's cost items are shown in *Figure 5-1*.

Costing of accidents indicates clearly the losses which have already occurred and may in some cases be continuing. The loss of trained workers from the workplace, the cost of replacing them, and the learning curve which applies to replacement workers, all impact negatively on production efficiencies. The cost of pensions and cash settlements will ultimately affect insurance premiums whether the insurer is the Workers' Compensation Board or a private carrier.

Following similar logic to combat the escalating costs of compensation, certainly in North America, early return-to-work programs are seen as offering some relief, together with alternate duty assignments for workers recovering from injuries. Unfortunately, all such approaches are in the "reactive mode", that is, an accident has taken place, an injury has occurred and actions are necessary to reduce the negative effects on the company's results. The readiness with which a modified work program for injured workers is embarked upon, rather than an overall system dedicated to

This chapter contributed by Don J. Pedley, Engineering Director, Loss Management Services, Industrial Accident Prevention Association, Ontario, Canada.

COMPENSATION AND BENEFITS COST

Workers' Compensation benefits payments:

- Medical and rehabilitation costs
- Pension and lump sum payments
- Wage replacement
- WCB administration costs

Additional company benefit costs:

- Accident insurance
- Death grant
- Life insurance
- Long term disability
- Other benefits
- Survivor(s) pension(s)
- Wage supplement/differential

LEGAL COSTS

Fees for external consul

- Trial and hearing costs
- Expert witness fees
- Fines
- Time for witnesses to attend

Financial settlements:

- Union grievance costs

TIME/PRODUCTIVITY AND PRODUCTION LOSSES

Marginal cost of product/production replacement

Employee time spent in post-accident activities:

- Cost of accompanying victim to hospital/home
- Time assisting/watching at time of accident
- Time by supervisors & upper management investigating, reviewing, etc.
- Time/costs of hiring replacement staff
- Time on clean up and salvage
- Time on laboratory analysis
- Time on repair of production equipment/site
- Time preparing investigation reports
- Time purchasing replacement materials/equipment

Productivity losses due to accident:

- Alternate transport costs
- Cost of learning curve of replacement workers
- Costs of make-up overtime
- Cost of reworking product
- Decreased output upon return to work
- Demurrage
- Increased maintenance costs
- Light duty costs
- Set-up and start-up cost
- Temporary slow down/stop-work of workers due to increased diligence/fear

MATERIALS, EQUIPMENT, AND PROPERTY DAMAGE

- Customer return: Poor product quality
- Equipment replacement capital cost
- Equipment replacement rental cost
- Property damage
- Process material product lost or damaged

MISCELLANEOUS OTHER LOSSES

- Consultant fees
- One time (knee jerk) safety costs
- Transport costs for victim

SUBJECTIVE LOSSES - TO BE NOTED BUT CANNOT BE COSTED

- Public relations
- Employee relations
- Corporate image
- Pain and suffering of victim
- Employee morale trauma
- Supervisor/manager trauma
- Potential lost customer/markets
- Potential contract bargaining demands

Figure 5-1

This chapter contributed by Don J. Pedley, Engineering Director, Loss Management Services, Industrial Accident Prevention Association, Ontario, Canada.

the prevention of accidents, is yet another example of the common perception of inevitability mentioned previously.

Again, to use the quality example: the inevitability of variable quality and the existence of repair departments at the end of each production line was very much a mater of common acceptance until the quality movement impacted North American manufacturers in the late 1970's and early 80's. From the "make it first - fix it later" philosophy which recognized the inevitability of less than perfect production, industries rapidly moved to the quality management requisites that demanded the establishment of an upstream control system. The resulting need for analysis, procedures and standards-setting has resulted in the wide adoption of measurement and rating systems.

The myth of the inevitability of rejects has been dispelled, but the myth of the inevitability of accidents and injuries remains, albeit temporarily. I say temporarily, because I am confident that once the full benefits of the quality movement have been recognized in producing products acceptable to the market place at an economic cost, the adopted principle of continuous improvement will cause organizations to recognize other fields where profits are slipping away. High on the list for consideration at that time must be the costs of accidents, not only injuries to persons but the costly consequences of other events caused by similar system failures or "management oversights", as W. G. Johnson, the designer of MORT, puts it.

If a company is experiencing accidents which injure persons, that fact will be obvious to the workers, the supervisors, and to the accounting and human resources departments. This will be even more glaring in the event that the insurance bill dramatically increases, or if an enforcement authority leaves citations which result in penal-

This chapter contributed by Don J. Pedley, Engineering Director, Loss Management Services, Industrial Accident Prevention Association, Ontario, Canada.

ties or fines. The experience will also likely reach the corner office and the boardroom. At that time, action may well be initiated to look into causes - but all too frequently with the production line "fix the defect" approach, rather than with a long-term plan in mind.

It is all in the establishment of priorities. In 1994 IAPA's more than 100,000 member firms reported 67,135 disabling claims. However, only 15,400 organizations reported any claims; 85% of members being claim-free. Ontario is primarily a Province of small businesses and 26.4% of the 1994 disabling claims occurred in the 189 member firms with over 1,000 employees. Member firms of over 250 employees reported, 45.1% of the total claims. It is not surprising, then, that the more than 87,000 firms without a claim in 1994 should see little or no need to address *accident* prevention in its narrow interpretation as *injury* claim-prevention.

But what of the wider impact on company profits of the accidental losses which *do not* result in injuries and compensation claims? Injuries are not the only consequence of accidents. Such research as exists indicates that incidents with damage to property, machinery and materials exceed those injuring people on a scale of three to one.

Here clearly is a cost to the enterprise that needs to be recognized and controlled. A skilled consultant visiting a manufacturing, processing, warehousing, or retail operation will quickly be able to determine the existence of this hidden cost. It is obvious from the paint scrapes on the lift trucks, the evidence of structural damage to buildings, the leaking sacks and damaged cans in the warehouse, and the repair welds to machinery and equipment, that ongoing damage is taking place. Is this inevitable? Of course not. But, as for the injury accident, symptom treatment is not sufficient to prevent recurrence. It is not enough to warn

This chapter contributed by Don J. Pedley, Engineering Director, Loss Management Services, Industrial Accident Prevention Association, Ontario, Canada.

or even re-instruct the responsible employee, in the event that one can be identified. It is not sufficient to issue general warnings or even post notices. What is needed is the recognition that, without adequate standards and procedures, and without the implementation of a loss control management system, there will continue to be damage, scrap, breakdowns, injuries, and other related losses.

Some companies have made great strides in the field of comprehensive loss control. Notable are the petroleum and chemical industries where the catastrophic nature of potential failures may have been the early incentive to implement prevention systems. In these, and a growing number of other sectors, the systems put in place for the control of both potential injuries and potential damage have served them in good stead from an overall business perspective. To most well organized, successful companies, such built-in controls go without saying.

Some years ago, the Toyota Company, which had recently set up an automobile plant in Cambridge, Ontario, appeared before a local safety conference audience with two speakers who told of their plans for employee safety. First an engineer from Japan told of the parent company's proud record and how it was accomplished. He spoke of safety posters, a safety arch under which workers entered the plant, a safety figure whose eyes were open or closed in proportion to the department safety performance, the buddy system, and many other behavioral techniques. This gentleman was followed on the podium by the newly appointed safety coordinator for Canada. He told of some parent company awareness programs which were being instituted for Canadian workers and, in addition, mentioned that the firm was registered in the International Safety Rating System TM.

This chapter contributed by Don J. Pedley, Engineering Director, Loss Management Services, Industrial Accident Prevention Association, Ontario, Canada.

He then drew audience attention to the program elements that Toyota would be working on. Among these were Organizational Rules, Employee Training, Personal Protective Equipment, Personal Communications, and Group Meetings. I was astonished. For a newly established work place, where were the key elements: Leadership and Administration, Management Training, Task Analysis and Procedures, Purchasing and Engineering controls? At the end of the session I approached the speakers in private and asked my question. They were as taken aback as I had been earlier. "Those things", they said, "are not a part of safety; they are a part of our basic management system".

So it is that, with very aware and successful companies, safety, health and loss control is not a special program, any more than Quality Control, Production Control, or even workplace-related environmental control. They are all part of the basic management system.

It is not by coincidence or someone's personal whim that, outside an entrance of one of Dupont's Ontario plants, there is a sign that records the number of days since the last "off-the-job" injury. Apart from the fact that Dupont Canada, in keeping the Dupont world performance, experiences very few "on-the-job" accidents, it was obviously recognized that the absence of a worker from his assigned responsibility, apart from all humanitarian consideration, is a cost-burden to the corporation. As such, the worker's overall health and safety is imperative, as is that of his or her family, in helping to reduce absences and also elevated stress levels. The fact that accidents off-the-job are seen as costly to families and to organizations as workplace accidents is but a small example from a corporation noted for its history of business success and also known for its world leading accident records.

This chapter contributed by Don J. Pedley, Engineering Director, Loss Management Services, Industrial Accident Prevention Association, Ontario, Canada.

A past senior vice-president of Dupont Canada, writing about the company's struggle to maintain its safety excellence in a period of transition toward self management, notes, "The best practitioners have moved the safety performance of their organizations far up the scale. Using the best safety technology and motivational concepts of self management and involvement, they are achieving safety results without additional cost—once they know how to do it. This does not mean that excellent safety performance causes excellent profits; rather, they can be concurrent results of the same management concepts and practices."

In the same vein, at a Project Minerva workshop/symposium held at Queen's University, Kingston, Ontario, in September, 1993, Mr. David Colcleugh, Senior Vice President, Dupont Canada Incorporated, posed the question, "Why do we in Dupont spend so much resource on preventing injury and illness?" He answered it with:, "We do it for our stakeholders for business and humane reasons, to improve our revenues and reduce our costs, keeping key people on the job, improving quality and reducing compensation, overtime costs, and so forth."

Specific research on the contribution of safety to profitability has been slight and generally inconclusive. However, research carried out in the late eighties by Dr. Larry Gaunt of Georgia State University, indicated that for companies reaching a high level of control as measured by the International Safety Rating System_, the use of that system had a strong positive effect on profitability. Similar research with ISRS users in Ontario indicated that 56% of respondents believed that the use of ISRS had brought about positive change, and 54.4% reported that corporate profitability was better than before.

Profitability is of course subject to many variables: the

This chapter contributed by Don J. Pedley, Engineering Director, Loss Management Services, Industrial Accident Prevention Association, Ontario, Canada.

company, competitive pressures, market share product management strategies, suppliers' material costs, labor peace, to name but a few. Similarly, the savings and potential profit deriving from a successful safety program will be subject to debate. It is relatively simple to account for the salaries of the safety department, the cost of personal protective equipment, the cost of safety incentives, literature, and training programs, but the offsetting and resulting benefits are not so easy to quantify. How does one measure that which has been prevented? To make the point, let us try some possible categories of savings.

1. Reduced compensation insurance premiums.
2. Reduced costs attendant on compensation insurance settlements, e.g., appeals, hearings, litigation.
3. Reduced productivity and production losses (as identified in the Hansen Model).
4. Reduced scrap and materials damage.
5. Reduced damage to equipment and tools.
6. Reduced damage to finished product.
7. Reduced property damage.

Some of the categories contain items which may be recognized but not categorized as accident related, and therefore not charged to a "lack of safety" account. To properly identify savings will therefore require a system of categories, the simplest of which is a two-part record-keeping by those responsible for replacement or repairs. This is the vital question: "Is this loss (cost) caused by an accident or by something else, for example, fair wear and tear, deliberate act, forces of nature, and so forth?"

A two-column system will enable an organization to address the nature of the loss, evaluate the basic causes, and put a

This chapter contributed by Don J. Pedley, Engineering Director, Loss Management Services, Industrial Accident Prevention Association, Ontario, Canada.

system or procedure into place to prevent reoccurrence. Examples might be as shown in *Figure 5-2*.

CAUSES		
Cost Factors	Accident	Other
Need for replacement workers caused by	Injured workers	Vacations
Need for overtime	Injured workers or equipment breakdown	Delivery schedule
Tool and equipment repair costs	Accidental damage	Wear & tear
Parts rework costs	Accident/set up error	Tool wear
Materials or finished goods damage	Accidental damage or spoilage	Vandalism

Figure 5-2

With the understanding that all accidents are preventable, such a record-keeping system will permit the recognition and costing of losses which downgrade the organization's performance. Even more importantly, however, the system serves to identify those accidents which should be fully investigated to determine the basic, underlying causes. This permits specific measures necessary for a long-term prevention system with consequent savings. The company that identifies and takes steps to eliminate its potentials for on-going losses is the company achieving control.

Naturally, the investment in this process must meet a normal business cost/benefit test. Considering the gross potential for savings within the full scope of loss control, this process is likely to pass the test and yield an early return on program investment.

In summary, the traditional role of management has been to plan, lead, organize and control. Within the planning phase falls the identification of business risks and potential

This chapter contributed by Don J. Pedley, Engineering Director, Loss Management Services, Industrial Accident Prevention Association, Ontario, Canada.

losses. Modern management has recognized that potential losses certainly include waste, scrap, production inefficiencies, and workers absent from assigned tasks. Lessons learned from leaders and workplace experience highlight the need to identify the basic underlying causes of losses potential losses. Consistently, these causes are found upstream in the management system.

In looking for greater efficiencies and lower costs, there is a clear opportunity for increased success by recognizing the costs of downgrading incidents and accidents. Control of those current and potential costs by the implementation of an excellent safety management system brings many benefits, including a positively improved BOTTOM LINE.

This chapter contributed by Don J. Pedley, Engineering Director, Loss Management Services, Industrial Accident Prevention Association, Ontario, Canada.

Selected References

Accident Prevention Manual for Industrial Operations. 4th Edition. Chicago: National Safety Council.,1959.

An Accident Costing Model for Use by Ontario Industry. IAPA, 1991.

Bird, Frank E., Jr. *Management Guide to Loss Control.* Loganville, GA: Institute Publishing, 1974.

Gaunt, Larry D. *The Effect of the International Safety Rating System (ISRS) on Organizational Performance.* College of Business Administration Georgia State University - Research Report 89-2.

HSE. *The Cost of Accidents At Work.* HMSO, 1993.

Harvey, Ted. Dr. SPR. *The International Safety Rating System, an Evaluation of Effectiveness.* IAPA. February, 1990.

Heinrich, H. W. *Industrial Accident Prevention, A Scientific Approach.* McGraw-Hill, 1931.

Johnson, W. G. *Management Oversight & Risk Tree.* MORT. US Government Printing Office, 1977.

Oxenburg, Morris. *Increasing Productivity & Profit Through Health & Safety.* CCH Australia Limited, 1991.

Presentation at a Project Management Workshop. Queen's University. Kingston, Ontario S.1, 1993.

Safety Subjects Bulletin No. 67. US Department of Labor. Bureau of Labor Standards. Washington, D.C: US Government Printing Office, 1955.

Simonds, Rollin H. *Estimating Costs of Industrial Accidents.* US Department of Labor. Washington, D.C: US Government Printing Office, 1955.

Stewart, J. V. Professor. *The Multi Ball Juggler.* Business Quarterly. Summer, 1993.

This chapter contributed by Don J. Pedley, Engineering Director, Loss Management Services, Industrial Accident Prevention Association, Ontario, Canada.

Chapter 6

MISSION ORIENTED SAFETY

"Not only is profit a measure of efficiency - good profit actually promotes efficiency by enabling a business to plan and invest soundly for the future."

Frederick R. Kappel, Former President, American Telephone & Telegraph

INTRODUCTION

Orienting safety to the mission of an organization today requires more than doing what we have done in the past to control accidents. We have not done well in demonstrating the huge contribution to be made to the mission of the organization. In addition to controlling accidental loss, we must demonstrate the other benefits safety has on the entire organizational system. Many safety professionals continue to try to manage a safety program in the same manner they did 25 years or more ago. As a result, many leaders do not see safety as making a positive contribution to the mission of an organization, or even producing the maximum value in accident control for which the organization pays. Many leaders at all levels apparently see safety activities as extra baggage they are told to carry, in addition to the heavy load of their other responsibilities. This reaction may be why so many program coordinators have not earned the management commitment they so avidly seek.

Significant insight to these harsh words can be gained from a key statement made by Martin P. Weinstock in, "The Business Advantage" (*Occupational Hazards*, June 1994). He was reporting his reactions to comments junior executives shared with many environmental, health and safety (EHS) professionals at a safety conference he had recently participated in. The relevant words are: "In an era of cutthroat competition, say some leading edge companies, the race is not just to the swift or strong, but to those who embrace environmental, health and safety as an essential part of their business." He continues, "speakers told the gathering of 200 EHS professionals that companies must stop treating EHS as an afterthought and integrate it into their everyday business operations."

We interpret this (and many other similar comments on the integration of safety into "business operations") to imply that many organizations are not doing it. The big question is, is this a failure of management or have we as safety leaders somewhere missed the mark?

We suggest that management is reacting in a normal way to a pill (EHS) that doesn't have the appealing taste that it ought to have.

Larry L. Hanson, CSP, ARM, presented a fresh message to

his safety colleagues in "Re-Braining Corporate Safety and Health," (*Professional Safety*, October, 1995). He points out that risk managers are questioning the wisdom of traditional safety philosophy and that the message is clear, "The future belongs to those willing to question current ways of doing things," and "re-braining safety requires a major shift in current beliefs about what drives safety performance."

The remainder of this chapter gives our opinions on "re-braining," that is, changes we believe are needed to drive safety performance in a way that will better demonstrate its mission-oriented value. It includes these four sections:

- Mission: The Bottom Line and More
- Successful Mission Oriented Changes
- Cultural Change Needed
- Action Step Guidelines.

MISSION: THE BOTTOM LINE AND MORE

Any reference to mission in the 1990's automatically causes a business leader to refer to his/her mission statement, since this may vary with each organization. No longer does the simplistic answer of the 50's concerning economic results or profitability suffice. You can recall a significant number of profitable firms who ceased operations in the last decade because of damage to their public image from an accident prone product. Image is only one of many mission-related concerns that any firm must protect in today's business climate.

The development of a suitable mission statement is an early step in an organization's strategic planning process. This accepted business practice is supported by many studies indicating that a business which engages in strategic planning, starting with the foundation of a firm mission statement, significantly outperforms those which do not.

John A. Pearce III, Ph.D. of George Mason University, and Fred David, Ph.D. of Auburn University, state the function of a mission statement in, "Corporate Mission Statement —The Bottom Line" (Academy of Management Executive, Vol. 1., No. 2):

> Mission Statement Function
>
> An effective mission statement defines the fundamental unique purpose that sets a business apart from others of its type and identifies the scope of the business operation in product and market terms. It is an enduring statement of purpose that reveals an organization's product or service, market, customers and philosophy.

Since this document could really be called a "statement of purpose" that defines the organization's business, it is the cornerstone around which management's operational strategies are built. In effect, it is the specification for the existence of the business.

Mission statements tend to focus on these eight organizational concerns:

Mission Statement Components

1. Key elements of philosophy
2. Organizations' self-control
3. Desired public image
4. Targeted customer market
5. Principle product or service
6. Geographical domain
7. Identification of core technologies
8. Commitment to survival, growth and profitability.

Studies have indicated that the average manager places most importance on number three (desired public image) and number eight (commitment to survival, growth and profitability). An exciting and welcomed phenomenon is the growth of EHS commitment as a major part of number three. Most television viewers can remember one or more chemical, oil or mining company's advertisement containing a statement of the organization's unique concern with EHS. Dow, DuPont, Proctor and Gamble, Sun Oil and Syncrude are among the industrial leaders visibly responding to the call of an ever growing group of EHS activists calling for much greater corporate responsibility to prevent a future Bhopal, Valdez and like disasters.

A long list of factors such as compliance with regulations, large fines, huge court awards, and imprisonment could be listed that, individually or in combination, are motivating this vital concern for mission. None of these factors is greater than the highly influential public activism of environmentalists, religious organizations, the usual industrial leaders who set an early pace, stockholders, board members and, perhaps of greatest importance, consumers. Executive stockholders and board members pay special attention to those things that adversely affect public image and product sales.

There is a growing rumor involving that the International Organization for Standardization will release a standard for safety management in 1999. "The ISO Man Cometh: Moving to Global Standards," referred to earlier in this book, is typical of the articles causing stockholders and board members to express deep concerns such as: Are we ready? How do we compare with world-class standards? Are we in compliance? Do all of our plants conform? Questions like these will be asked more and more by those with a vested interest in successfully competing in the world market. The result of this growing activism of companies to "clean up their EHS Act and get with it," will be reflected in more and more company mission statements. Robert H. Hayes, Professor of Business Administration at the Harvard Business School, specializing in manufacturing strategy and international competitiveness, and Gary P. Prisano, Associate Professor at the Harvard Business School, specializing in technology and manufacturing strategy, provide some very interesting thoughts in, "Beyond World Class: The New Manufacturing Strategy" (*Harvard Business Review*, January - February 1994). Their futuristic thinking regarding management strategies planning centers about this significant thought,

> "How can an organization expect to achieve competitive advantages if its goal is to be as good as the toughest competitor? To be competitive and lead - our goal must be - Beyond world-class".

Could the influence of these great strategists significantly increase management commitment to safety? Quite likely. Especially when other influential colleagues like Dr. Quinn Mills, also of the Harvard Business School, is often quoted from one of his many published articles: "The only substantial advantage that any organization is going to have is the ability of its people."

Could this leading edge thinking of those who are influencing business leaders be one of the reasons more and more executives are committing themselves to becoming the best in mission statements such as this:

Environmental, Health and Safety Excellence

We are fully committed to safety and environmental protection. We are responsible to our employees and those persons in surrounding communities. Production or profit will not compromise the safety of our people or protection of the environment. **We will be second to none in our industry.**

Successful Mission Oriented Changes

The following actions contribute to the mission of a company and are an official part of the safety activities of thousands of leading organizations. These actions are the result of a systematic approach integrating safety into every possible part of the organizational system. They are not listed in any order of priority.

1. Systematic identification of critical tasks, that is, those that if done improperly could result in a major safety, quality or production loss.

2. Systematic identification of conditions such as leaks, uncontrolled overflow, waste disposal, etc. that could result in costly environmental contamination of the internal or external plant soil. This is accomplished in the regular inspection for historically identified safety hazards.

   Housekeeping inspections tend to focus on substandard safety, health and environmental conditions. However, good housekeeping (cleanliness and order) is also essential to quality, productivity, cost control, morale and company image. Good housekeeping is a critical aspect of mission oriented safety.

   It has frequently been said that nothing impresses a customer more about the quality of your product than the appearance of your shop floor, where the product is produced. This appearance is a direct result of compliance with the housekeeping standards that are a cornerstone of a good safety program. Its relationship to environmental management is shown in *Figure 6-1.*

"Clean Up the Shop"

Good Housekeeping Practices with Big Impacts on Emissions

1. Conduct regular inspections to ensure that equipment has no leaks and is operating properly.
2. Check flange joints
3. Replace leaking valves and flanges
4. Close dampers or doors during operation of equipment.
5. Ensure that ventilation equipment is operating effectively.
6. Inspect oil/water separators for evidence of contaminated discharges.
7. Inspect interior floor drains for evidence of contaminated discharges.
8. Make certain that all drums are labeled.
9. Discard oily rags in a disposal drum.
10. Inspect drum storage areas for leaks
11. Clean up minor spills immediately.
12. Do not overfill fuel
13. Ensure the MSDSs are easily accessible by employees.
14. Observe and report and unusual workplace odors
15. Report any unusual sounds in the workplace.
16. Keep the workplace clean and orderly.

It is important to note that maximum available pollution control standards (MACT) include work practices or operational standards including training or certification of operators to reduce the emissions of hazardous air pollutants. These can often be implemented quickly and cheaply. It is known, for example, that closing dampers or doors when certain equipment is in operation is sometimes sufficient to eliminate emissions without the use of enhanced control techniques.

Figure 6-1

3. Expansion of the regular safety inspection process to include reporting any unusual condition or circumstance that could impede effective quality or production operations. Emphasis is on predictive maintenance and equipment that will result in major loss if not operating properly.

4. Expansion of "job safety analysis" to integrated (safety-quality-production) critical task analysis (*Figure 6-2*). Critical tasks are analyzed step by step with

CRITICAL TASK ANALYSIS WORKSHEET

Concentrator	Start Particle Size Monitor	W. Livingstone Eng. 5/18/19–
Plant/Division	Critical Task Analyzed	Signature Function Date
Grinding	5/14/19–	R. Hagan Supv. 5/16/19–
Department	Date Completed/Revised	Signature Function Date
Grinding Operator	E. Bromley	R. Swinehart Supv. 5/23/19–
Occupation	Initial Approval	Signature Function Date

No.	Task Steps	Specific Loss Exposures (Safety-Quality—Production)	Recommended Controls
1	Record each gauge's current reading (Particle Size Monitor)	Equipment damage, poor readings	Follow checklist
2	Look for sand buildup in cyclone box	Equipment failure will force coarse particles into system	Clean any accumulation
3	Hose out air eliminator	Damage to air eliminator from sand	Clean any accumulation
4	Close drain valve	Water spills, electric shock	
5	Open fresh water valve	Inadequate flow of water will cause plug-ups	Clear water filter
6	Clear sensor pump	Hand injury from wrench slipping	Shut down before cleaning, use special cleaning tool
7	Check placement of main sensor	Poor reading	Replace if missing
8	Put safety system on automatic	Damage from low water pressure	
9	Clear sample screens in cyclone overflow box	Poor readings	Visually inspect. Clean any accumulation
10	Pull start button		
11	Raise water pressure to sixty (60) psi	Automatic shut down of system	Record water pressure gauge reading every thirty (30) minutes
12	Fill tank	Overflow, clogs, system shutdown	Listen for safety water to cycle off and on
13	Place sample screen in slurry	Inadequate vacuum	Look for normal vacuum of 17+ inches
14	Set intake flow valve at position five (5) for normal slurry level	Overflow or inadequate flow of slurry. slips and falls to surface	Allow 5 minutes to stabilize

Figure 6-2

the worker performing the task in order to find the most efficient way to do these tasks. When a more efficient way than the established way is suggested or observed, the suggested improvement, with expected efficiencies, is submitted to management for approval as part of the right way (not just the safe way) to do the job.

5. Expansion from "safety procedures" to "standard task procedures." Establish written procedures that cover the safe, high quality, productive way to perform each critical task (*Figure 6-3*). This not only produces better procedures, but also reduces the number of procedures that must be produced and maintained.

6. Update critical task procedures whenever a change takes place, whenever any problematic event occurs, whenever a better way to do the task is found, and at least, within an established period of time (that is, annually) for each task. This regular updating becomes a leadership habit and is one of the best tools enabling continuous improvement.

7. Observe critical task performance at established frequencies to determine if a better way has been found to complete the task. If so, update the procedure and coach employees on the improved procedure. Note that, in a system oriented safety program, planned task observations are geared to the right way to perform the task (not simply the safe way).

8. "Key point tipping" is required and taught as part of a modern, mission oriented safety program. Key points are those vital bits of information that make or break the job; special tricks of the trade that make the task more efficient; the feel or knack that is the mark of a pro; critical points of quality, productivity, cost control

Standard Task Procedure

ISSUED TO:

Start Particle Size Monitor — TASK

June 20, 19— — DATE OF ISSUE

Grinding — DEPARTMENT

Grinding Opr. — OCCUPATION

REVIEW DATES

TASK PURPOSE AND IMPORTANCE

The particle size monitor analyzes the grind and density of the final grinding product. The trends are recorded on chart recorders. The result is a record of representative samples of the total day's grind.

The proper adjustment of the entire grinding and flotation process is dependent upon accurate sampling techniques. Errors and malfunctions can be responsible for many thousands of dollars of wasted ore, materials and efforts. Some of the equipment involved is easily damaged if not properly operated and regularly inspected.

The major steps are outlined in their proper order. Key points to remember follow each step. All steps and key points must be followed in sequence to achieve maximum efficiency and avoid losses.

1. Record all instrument panel gauge readings. Use check list.
2. Look for sand buildup in cyclone box and clean any accumulation before startup to avoid equipment damage and poor readings.
3. Hose out the air eliminator.
4. Close drain valve to avoid spillage and possible falls or electric shock.
5. Open main fresh water valve and clear water filter to assure adequate flow of water.
6. Shut down and clear sensor pumps. Use special cleaning tool to avoid disassembly of pump and possible hand injury.
7. Ensure that sensor is in place.
8. Put water system safety switch on automatic to avoid damage from low water pressure.
9. Visually inspect the sample screen in the cyclone overflow box and clean any accumulation to avoid poor readings.
10. Pull start/stop button on control box to start.
11. Ensure that water pressure is at least 60 psi to avoid automatic shutdown of system and record reading every 30 minutes to ensure that this pressure is maintain.
12. Allow tank to fill until the sensor section is full, at which point the safety water will automatically turn off. Listen for safety water to cycle on and off.
13. Place the sample screen in the slurry. The vacuum should rise to normal (17+ inches). Visually check.
14. Adjust sample intake flow valve for normal tank slurry level - allow 5 minutes to stabilize to avoid overflow or starving the system.

STANDARD TASK PROCEDURE

I have received the Standard Task Procedure on this date. It has been explained to me and the importance of following it has been stressed.

Received by: ____________________

Date: ____________ Supervisor: ____________________

Figure 6-3

or safety. "Tip" means just what it does in everyday life: a piece of information given in an attempt to be helpful; a small gift, a hint or suggestion. Key point tipping, then, is the organized process of giving employees helpful hints, suggestions, reminders or tips about key quality, production, cost or safety points in their work.

A tip can be as simple as:

"Jerry, remember to tie in the top of that ladder. I'd hate to see you fall and hurt yourself."

Or

"On this set-up, Sandra, be sure to follow the pre-operation checklist. It will save you a lot of time and rework."

Supervisors/team leaders are taught to, "give a key point tip with every critical task assignment." Safety, quality and productivity are also integrated in proper task instruction (PTI) as shown in *Figure 6-4.* A proven approach to giving effective talks is the "5P" method shown in *Figure 6-5.* It is useful for talks on safety, environment and quality topics.

9. Mission oriented leaders apply behavior reinforcement not only to safe behavior, but also to quality behavior and productive behavior, that is, to all behaviors desired for meeting the mission. When they notice someone doing something particularly well or demonstrating significant improvement, they reinforce (strengthen) that behavior by providing consequences. When they observe substandard behavior, they provide constructive coaching. A performance management approach such as shown in *Figure 6-6* may be used to best decide what type of corrective action

is most appropriate to the situation.

10. The incident investigation form and procedure is so designed that it permits any inadequately controlled event, whether involving safety, quality or productivity, to be adequately investigated.

Safety and Quality through Proper Task Instruction

Supervisory Responsibility

Proper job instruction "PTI," must be given with every task or every new task. This instruction must include all necessary phases of the task done in order to insure safety, quality, productivity and delivery on time. PTI is no a complicated procedure nor does it take a lot of extra time or talk. On the contrary — energy and materials are saved and pain spared with PTI has been given at the right time — before the worker starts the task — not after it was started and has done something the wrong way to expose himself to injury, cause a product rejection or costly delay. In order to present PTI properly, you should understand fully the two basic steps involved:

STEP 1—Tell the worker the importance of doing this task properly.
STEP 2—Make sure he or she knows how to do it properly by:

A. TELLING	Tell the worker clearly how to do the job correctly. Tell him/her clearly how to ensure quality and completion on time. On most occasions, only a few words should be necessary, however, the detail of PTI will depend upon the work being assigned. We can't expect a job to be done safely and with desired quality unless we have first given proper instructions for the job.
B. SHOWING	When certain jobs are being assigned, it is necessary to show the worker how to do the job properly as well as to tell him how to do it. The employee will better understand and remember the procedure longer if you personally have shown what you expect.
C. TESTING	The only way you can be sure that the worker understands what you have said, and even shown him, is to test his or her understanding. This is the ideal time to correct anything not understood. Remember the words ever available for this purpose - who? - which? - why? - what? - where? - and how?
D. CHECKING	(Or follow-up). This is probably the most important step in PTI. With it we get compliance. It is through your daily and hour-to-hour checking that you teach the worker who is forgetful and needs to be watched, spot the indifferent employee who takes safety or quality half-heartedly, or apprehend the few with the wrong attitude who deliberately disobey instructions.

Figure 6-4

How to Give Better Talks

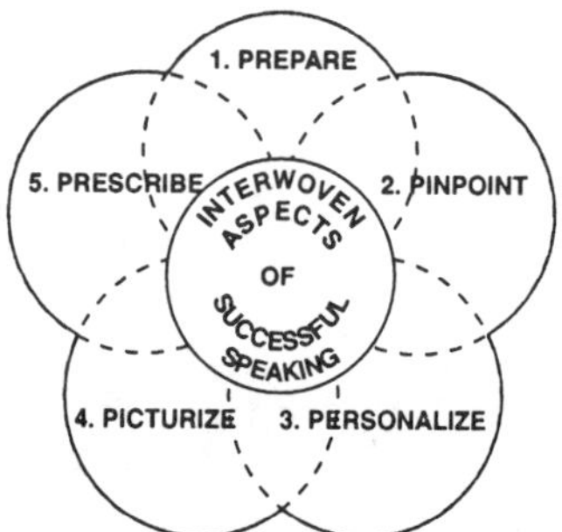

using the 5 Ps

PREPARING

- Think about the subject
- Write things down . . . for your idea-bank
- Read related materials selfishly
- Listen to other's ideas and attitudes
- Organize and outline your talks
- Practice

PINPOINTING

- Don't try to cover too much ground
- Zero-in on one main idea . . . that you can state in a single sentence

PERSONALIZING

- Establish common ground with your listeners
- Bring it close to home
- Make it important in their minds
- Make it personal and meaningful to them

PICTURIZING

- Create clear mental pictures for your listeners
- Appeal to both their ears and their eyes
- Help them to really "see what you mean"
- Use visual aids

PRESCRIBING

- In closing your talk, answer the question the listeners always have: "So what?"
- Tell them what to do
- Ask for special action
- Give a prescription

Figure 6-5

The incident investigation procedure is also designed to teach modern concepts of problem solving. Adequately designed, this exercise becomes a regular review of problem solving to use whenever an unintended event causing loss of any resources occur.

11. Operators of mobile materials handling equipment are licensed and trained in the most current methods to prevent loss of people, equipment or product.

12. Leaders learn the principle of the four-step technique

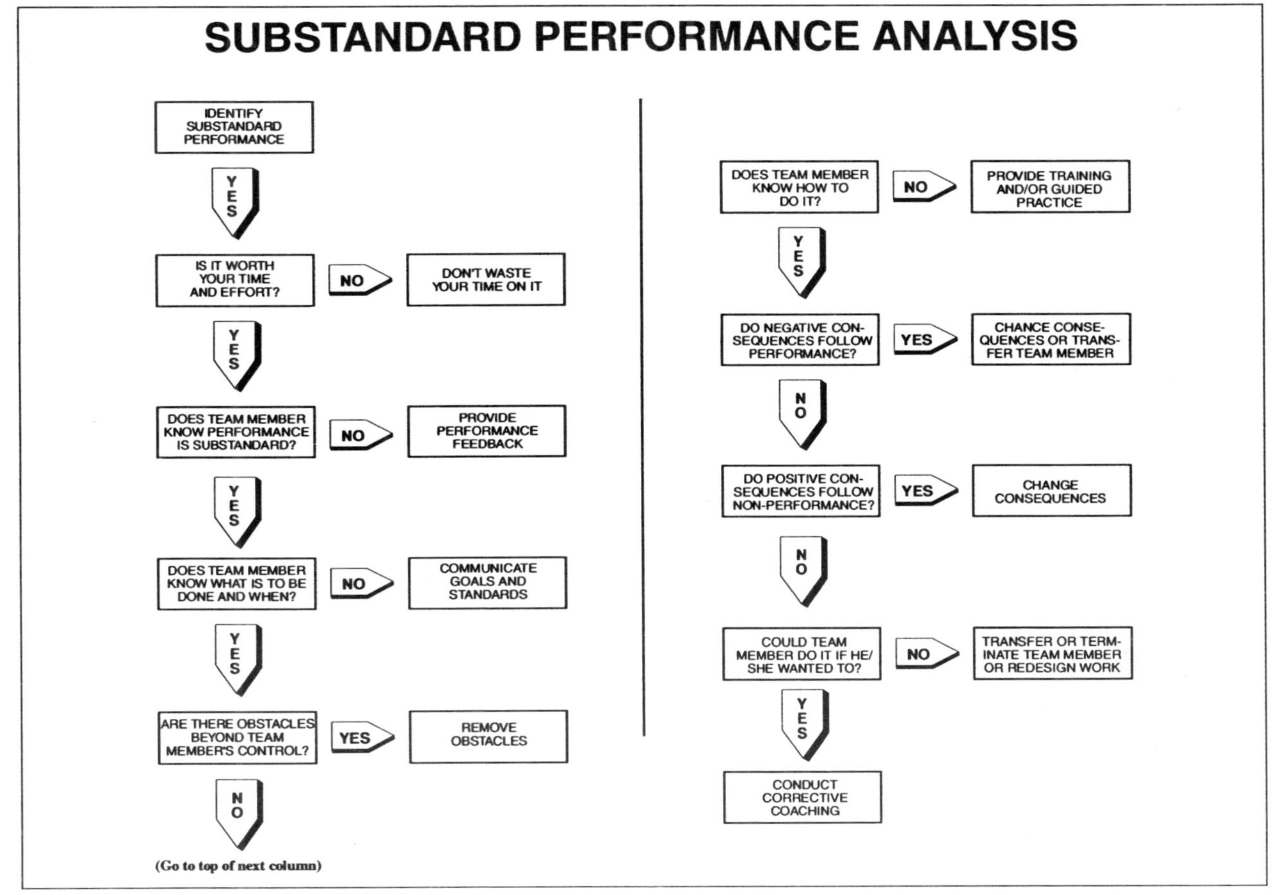
SUBSTANDARD PERFORMANCE ANALYSIS
IDENTIFY SUBSTANDARD PERFORMANCE
YES
IS IT WORTH YOUR TIME AND EFFORT?
NO
DON'T WASTE YOUR TIME ON IT
YES
DOES TEAM MEMBER KNOW PERFORMANCE IS SUBSTANDARD?
NO
PROVIDE PERFORMANCE FEEDBACK
YES
DOES TEAM MEMBER KNOW WHAT IS TO BE DONE AND WHEN?
NO
COMMUNICATE GOALS AND STANDARDS
YES
ARE THERE OBSTACLES BEYOND TEAM MEMBER'S CONTROL?
YES
REMOVE OBSTACLES
NO
(Go to top of next column)
DOES TEAM MEMBER KNOW HOW TO DO IT?
NO
PROVIDE TRAINING AND/OR GUIDED PRACTICE
YES
DO NEGATIVE CON-SEQUENCES FOLLOW PERFORMANCE?
YES
CHANCE CONSE-QUENCES OR TRANS-FER TEAM MEMBER
NO
DO POSITIVE CON-SEQUENCES FOLLOW NON-PERFORMANCE?
YES
CHANGE CONSEQUENCES
NO
COULD TEAM MEMBER DO IT IF HE/SHE WANTED TO?
NO
TRANSFER OR TERM-INATE TEAM MEMBER OR REDESIGN WORK
YES
CONDUCT CORRECTIVE COACHING

Figure 6-6

for proper task instruction (PTI) to use in all their leadership instruction activities. This is presented in many companies as safe job instruction. In many organizations with mission oriented programs. PJI is presented as one of the best ways to make sure any job or task is performed the right safe way the first time. The four step method to accomplish this is reviewed over and over again as a major step in preventing unintended accident or quality loss.

13. An effective way to present group verbal communications (not limited to safety) is taught to all leaders and quite frequently reviewed.

14. An opportunity is afforded to question or provide feedback on any important, timely issue through the vehicle of the regular safety meeting. Current psychological studies indicate that feedback is one of an organization's greatest motivational tool for developing employee loyalty, support and pride in their company.

15. Special theme programs are frequently designed to promote interest in safety, quality and productivity. Examples of such promotion efforts are shown in *Figure 6-7*. You can see that all of them mention not just safety, but also quality and production.

16. A modern safety program using the integrated system approach prepares an organization for the quality ISO 9000 certification, and/or the future ISO safety standards. Client organizations report their safety programs gave them a 70-75% accomplishment rate on ISO 9000 requirements.

17. An efficient safety program is evidence to customers, employees, outside inspectors, insurance companies, the public and others that the company cares about

ANTICIPATION

OBJECTIVES	KEY POINTS
a. To motivate increased preventive actions by teaching the meaning of anticipation.	a. Think ahead for safety
b. To create an awareness of the potential or problems by not anticipating.	b. Take safety actions when they are in order.
	c. Encourage others to anticipate, Anticipation Aids Safety

WHAT MAKES THE DIFFERENCE?

OBJECTIVES	KEY POINTS
a. Create an awareness of the big difference a little extra effort can make in safety, quality and production.	a. Just a little extra attention
b. To motivate everyone to a little extra effort in all areas of work.	b. Just a little extra thought.
	c. Just a little extra effort.

VIM (Very Important Motivator)

OBJECTIVES	KEY POINTS
a. To create an awareness of the influence that everyone has on others.	a. Be enthusiastic for safety
b. To stimulate everyone to be a very important motivator for safety, quality and productivity.	b. Be persistent for these areas of great value.
	c. Give recognition to associates for proper performance.
	d. Be a model of proper behavior yourself.

OPERATION ORDER

OBJECTIVES	KEY POINTS
a. To teach everyone the meaning of order and its important influences	a. Order increases efficiency.
b. To motivate everyone to practice order continuously.	b. Order creates a safety and quality-prone environment.
	c. Order increases productivity.

WHAT MAKES MISTAKES?

OBJECTIVES	KEY POINTS
a. To create an awareness of the most important reasons for the mistakes that cause accidents, production delays and quality defects.	a. Lack of knowledge.
b. To teach workers what to do when they become aware of personal deficiency that could be their mistakes.	b. Lack of skill.
	c. Improper attitudes.

DO YOU CARE ENOUGH?

OBJECTIVES	KEY POINTS
a. To motivate workers to realize how important their personal efforts and examples are in influencing others to do quality work safely.	a. To work safely for their own sake.
b. To create awareness within workers that their efforts and examples make a valuable contribution to the entire work performance of others.	b. To work safely for the sake of others?
	c. To set a safe example?
	d. to share safety with others?

Figure 6-7

their people. This is vital to good employee relations and corporate image. It reduces gripes, grievances, customer complaints and negative press.

18. An efficient safety program reduces lost time and absenteeism. Many studies have shown that employees want to work in a place that is pleasant, safe and has a caring culture.

19. An efficient safety program maintains compliance with the law and minimizes the potential of big fines.

20. A properly documented safe and healthy work environment can be used by lawyers as evidence liability cases to demonstrate legislative compliance and responsible care.

21. A modern safety program controls accidents that result in property damage. Accidental damage has been reported to be from 10 - 30% of maintenance costs and sometimes enough to double the bottom-line profit, if controlled.

22. An effective environmental, health, and safety system is a major contributor to building job pride with all its values to quality and productivity as well as EHS.

23. A modern safety system reduces potential errors that cause major and disastrous accidents as well many major quality or productivity losses for which the basic causes and controls are the same. The basic causes shown in *Figure 6-8* are valid for safety, quality and environmental loss events.

24. The skills consistently and persistently taught to front line leaders for safety motivation are the same skills used to motivate quality and productivity. The leaders learn to motivate people toward safe, high quality, productive, cost-effective behavior.

BASIC CAUSES - PERSONAL FACTORS

- **Inadequate Physical/Physiological Capability**
  - inappropriate height, weight, size, strength, reach, etc.
  - restricted range of body movement
  - limited ability to sustain body positions
  - substance sensitivities or allergies
  - sensitivities to sensory extremes (temperature, sound, etc.)
  - vision deficiency
  - hearing deficiency
  - other sensory deficiency (touch, taste, smell, balance)
  - respiratory incapacity
  - other permanent physical disabilities
  - temporary disabilities
- **Inadequate Mental/Psychological Capability**
  - fears and phobias
  - emotional disturbance
  - mental illness
  - intelligence level
  - inability to comprehend
  - poor judgment
  - poor coordination
  - slow reaction time
  - low mechanical aptitude
  - low learning aptitude
  - memory failure
- **Physical or Physiological Stress**
  - injury or illness
  - fatigue due to task load or duration
  - fatigue due to lack of rest
  - fatigue due to sensory overload
  - exposure to health hazards
  - exposure to temperature extremes
  - oxygen deficiency
  - atmospheric pressure variation
  - constrained movement
  - blood sugar insufficiency
  - drugs
  - Mental or Psychological Stress
  - emotional overload
  - fatigue due to mental task load or speed
  - extreme judgment/decision demands
  - routine, monotony, demand for uneventful vigilance
  - extreme concentration/perception demands
  - "meaningless" or "degrading" activities
  - confusing directions
  - conflicting demands
  - preoccupation with problems
  - frustration
  - mental illness
- **Lack of Knowledge**
  - lack of experience
  - inadequate orientation
  - inadequate initial training
  - inadequate update training
  - misunderstood directions
- **Lack of Skill**
  - inadequate initial instruction
  - inadequate practice
  - infrequent performance
  - lack of coaching
- **Improper Motivation**
  - improper performance is rewarding
  - proper performance is punishing
  - lack of incentives
  - excessive frustration
  - inappropriate aggression
  - improper attempt to save time or effort
  - improper attempt to avoid discomfort
  - improper attempt to gain attention
  - inappropriate peer pressure
  - improper supervisory example
  - inadequate performance feedback
  - inadequate reinforcement of proper behavior
  - improper production incentives

Figure 6-8

BASIC CAUSES - JOB FACTORS

- **Inadequate Leadership and/or Supervision**
  - unclear or conflicting reporting relationships
  - unclear or conflicting assignment of responsibility
  - improper or insufficient delegation
  - giving inadequate policy, procedure, practices or guidelines
  - giving objectives, goals or standards that conflict
  - inadequate work planning or programming
  - inadequate instructions, orientation and/or training
  - providing inadequate reference documents, directives and guidance publications
  - inadequate identification and evaluation of loss exposures
  - lack of supervisory/management job knowledge
  - inadequate matching of individual qualifications and job/task requirements
  - inadequate performance measurement and evaluation
  - inadequate or incorrect performance feedback
- **Inadequate Engineering**
  - inadequate assessment of loss exposures
  - inadequate consideration of human factors/ergonomics
  - inadequate standards, specifications and/or design criteria
  - inadequate monitoring of construction
  - inadequate assessment of operational readiness
  - inadequate monitoring of initial operation
  - inadequate evaluation of changes
- **Inadequate Purchasing**
  - inadequate specifications on requisitions
  - inadequate research on materials/equipment
  - inadequate specifications to vendors
  - inadequate mode or route of shipment
  - inadequate receiving inspection and acceptance
  - inadequate communication of safety and health data
  - improper handling of materials
  - improper storage of materials
  - improper transporting of materials
  - inadequate identification of hazardous items
  - improper salvage and/or waste disposal
- **Inadequate Maintenance**
  - inadequate preventive
    - ... assessment of needs
    - ... lubrication and servicing
    - ... adjustment/assembly
    - ... cleaning or resurfacing
    - ... inadequate reparative
    - ... communication of needs
    - ... scheduling of work
    - ... examination of units
    - ... part substitution
- **Inadequate Tools and Equipment**
  - ... inadequate assessment of needs and risks
  - ... inadequate human factors/ergonomics considerations
  - ... inadequate standards or specifications
  - ... inadequate availability
  - ... inadequate adjustment/repair/maintenance
  - ... inadequate salvage and reclamation
  - ... inadequate removal and replacement of unsuitable items
- **Inadequate Work Standards**
  - inadequate development of standards
    - ... inventory and evaluation of exposures and needs
    - ... coordination with process design
    - ... employee involvement
    - ... inconsistent standards/procedures/rules
  - inadequate communication of standards
    - ... publication
    - ... distribution
    - ... translation to appropriate languages
    - ... reinforcing with signs, color codes and job aids
  - inadequate maintenance of standards
    - ... tracking of work flow
    - ... updating
    - ... monitoring use of standards/ procedures/rules
- **Wear and Tear**
  - inadequate planning of use
  - improper extension of service life
  - inadequate inspection and/or monitoring
  - improper loading or rate of use
  - inadequate maintenance
  - use by unqualified or untrained people
  - use for wrong purpose
- **Abuse or Misuse**
  - condoned by supervision - intentional/ unintentional
  - not condoned by supervision - intentional/ unintentional

Figure 6-8 continued

The twenty four examples listed above typify the kinds of contributions to a company's mission that are made by a modern safety system. Such a system effectively maintains control of hazards that cause physical harm and property damage that can easily represent 25-50% of a company's net profit before taxes.

ACCIDENT COST IS PROFIT LOSS

Here is the bottom-line impact: When you decrease accidental loss, you increase profitability.

CULTURAL CHANGE NEEDED

Executive leaders determine the value and amount of emphasis the organization gives to safety, health and environmental protection. This "company climate" strongly influences how employees think, feel and act regarding safety. It is concerning this phenomenon that "the cultural change needed" comments are made. They certainly do not apply to that group of progressive business leaders like Ralph Sheppard, formerly CEO of Syncrude, Canada, who fully recognize the great contribution made by these vital subjects to the protection of all resources of an organization including the very life of its people as well as that of the business itself.

Unfortunately, too many leaders give what is historically referred to as "lip service" rather than the actual performance that would negate the words so frequently in the minds and hearts of those involved:

WE HEAR WHAT YOU SAY BUT WE SEE WHAT YOU MEAN.

Most executives would be more committed to safety if all

safety professionals presented the safety message in way that more clearly demonstrated its commonalties and contributions to the company's mission. We definitely agree with Douglas Clark and Per Olaf Brett, from their article, "Loss Control Creates Value and Competitive Advantage" (Forum, No.1, 1995), that the barriers which keep CEO's from taking the safety commitment ball more frequently and running with it are:

1. They don't appreciate the magnitude of the costs and wastes involved,
2. They do not believe there are any practical means of controlling risks,
3. They perceive loss control activities as expense items, necessary only to meet government regulations and protect employees.

As Pogo once observed, "We have met the enemy and it was us!" In the present context:

> The majority of safety leaders present safety as an add on activity to the management system. We need to present it as the value driven catalyst it can be when effectively integrated into the management mission and system.

This stimulates strong executive appreciation of its total contribution to the business performance and competitiveness advantage of the organization.

Allan T. Goldberg expressed it a bit differently in his, "Moving Past the Blame Basis: Safety as a Business Asset" (*Professional Safety*, January, 1996), "When safety is built into company policies, instead of added on, the level of protection is optimized, as is an organization's ability to flourish in a global economy."

The organization that misses this path to success is dreaming in the past and revealing a critical need to update its mission to meet today's and future competitive needs. It also needs awakening from its slumber to realize how much it has lost touch with modern cultural needs (including safety as a contributing value). A cultural change is indicated as a big need to provide the written and unwritten norms and values that provide the roadmap for mission achievement.

"Culture" in the 70's and early 80's was a popular buzz word for those things in an organization (knowledge and skill of employees, organizational strategy, incentives, etc.) that collectively gave it a kind of state of mind (how you do things around here) which distinguished it from other companies.

Since the early 80's an intensive search continues to determine those things that most influence an employee's desire to do quality work. By showing management how mission oriented safety activities really are, safety can earn its proper place along side of quality and productivity as the never ending search continues. None-the-less, the fierce competition which will continue to grow worldwide gives daily emphasis to this search.

Leading research psychologists and sociologists have pointed out that most evidence regarding employee motivation to achieve the desired level in excellence keeps pointing back to the organization's culture itself. Because of this, a vital management question continues to be : "What should we specifically do that we are not doing"? Research results point out that written and formal rules, procedures, policies and norms provide the powerful guideposts to mission achievement. But another even more powerful motivating force is always present. As previously stated, the components (influences) of an organization's culture are both written and unwritten (beliefs and expectations shared). The informal and unwritten norms are the part of the road map to achieving of excellence that we do not pay enough attention to. Its what leaders feel and think that may be considerably different from what they express in formal written directions; What psychologists and sociologists say most influence people in work performance; Causes the individuals whose desire we want to influence to say silently to themselves, "I read what you say you want, but I hear and see what you mean."

Here are specific guidelines for gaining management commitment through a mission oriented program (in effect, the mission of this chapter).

CONCLUSION

Action Step Guidelines

Here are specific guidelines for gaining management commitment through a mission oriented program (in effect, the mission of this chapter).

1. Strive to integrate as many activities of the safety process as possible into every facet of the management system.

2. Clearly show the value driven benefits of safety activities that link people and resource protection to business performance and company mission.

3. Do it again and again and again until the mission direction of safety is clearly perceived, understood and effectively implemented by management in all formal, written and unwritten norms and values for "the ways things are done" in your organization.

4. Celebrate success and communicate the details of your success to others.

Selected References

Brett, Per Olaf and Clark, Douglas, "Loss Control Creates Value and Competitive Advantage, *DNV Forum*, No. 1, 1995.

Goldberg, Allan T., "Moving Past the Blame Basis: Safety as a Business Asset, *Professional Safety*, January, 1996.

Hanson, Larry L., "Retraining Corporate Safety and Health", *Professional Safety*, October, 1995.

Hayes, Robert H. and Gary P. Prisano, "Beyond World Class - The New Manufacturing Strategy," *Harvard Business Review*, February, 1994.

Pierce, John A., III and George Mason, "Corporate Mission Statement - The Bottom Line," *Academy of Management*,Vol. 1, No. 2, 19.

Weinstock, Martin P., "The Business Advantage", *Occupational Hazards*, June, 1994.

Chapter 7

BEYOND BEHAVIORISM TO HOLISTIC MOTIVATION

PSYCHOLOGICAL PREMISE

During the late 1980s and early 1990s many safety consultants, authors and practitioners put major emphasis on the *behavioral side of safety, on managing behavior*, on skill training for specific *behaviors*. Some even carried it to the extreme, implying that behaviorism is the only sensible approach to performance management and continuous improvement.

At the same time, a major proportion of general management consultants, authors and leading practitioners emphasized concepts such as culture - vision mission - values - beliefs - trust - teamwork - attitudes - perception surveys. This shows that modern managers are concerned with

- "mental pictures" of the future
- goals - objectives - aspirations
- people's perceptions of "what is," "what can be," and

This chapter contributed by George L. Germain. MA, CHCM, ASA, Executive Consultant for DNV-Loss Control Management, Loganville, GA.

how to close the gap

- the organizational "climate"
- how people "feel" about things
- how to stimulate right-brain activity — innovation — creativity
- how to replace negative conflict and confrontation with positive cooperation and collaboration
- how to develop and maintain "loyalty" and "commitment"
- moral/ethical implications of management activities
- how to deal with the morale challenges created by the mind-bending pace of change, by the devastation of downsizing, by the rigors of reengineering

Some even carried this to the extreme of sadly neglecting the actual *work* (that is, the actions, the activities, the behaviors) required to satisfy their organizational and psychosocial concerns.

Neither extreme is desirable. A holistic approach to motivation and performance management is better. This incorporates the best of both: the behavioral and the psychological or psychosocial. Holistic motivation recognizes and utilizes the C-A-B framework used for decades by scientists and practitioners in the profession of psychology, especially regarding the teaching/learning process. This approach emphasizes the following vital subsystems:

C = Cognitive
A = Affective
B = Behavioral.

The *cognitive* area is what most people call "mental". It has to do with the mind mental pictures — concepts — strategies — thoughts — ideas — facts — information —knowl-

This chapter contributed by George L. Germain. MA, CHCM, ASA, Executive Consultant for DNV-Loss Control Management, Loganville, GA.

edge. The *affective* area is what most people call "emotional." It has to do with feelings — emotions — beliefs — values — convictions — mood — morale — disposition — attitudes. The *behavioral* area is what most people call "physical". It has to do with physical capabilities — actions — demeanor — dexterity — performance"hands-on" activity — skills.

As shown in *Figure 7-1,* knowledge, attitudes and skills are terms commonly used for these three vital avenues to human motivation and safety success.

Holistic Motivation

ATTITUDES
. Feelings
. Emotions
. Beliefs
. Values
. Conviction

KNOWLEDGE
. Information
. Ideas
. Thoughts
. Concepts
. Mental pictures

SAFETY SUCCESS

SKILLS
. Words
. Actions
. Behavior
. Capabilities
. Performance

Figure 7-1

However, the actual words we use are not the most important point. We may use terms like:

Cognitive - Affective - Behavioral

or

"Head" - "Heart" - "Hands"

or

This chapter contributed by George L. Germain. MA, CHCM, ASA, Executive Consultant for DNV-Loss Control Management, Loganville, GA.

Thinking - Feeling- Doing
or
Knowledge - Attitudes - Skills

The most important point is that effective motivation is holistic; it involves all three domains.

PRACTICAL PRINCIPLES

Figure 7-2 shows six practical principles for managing motivation. Let's look at a few key points regarding each principle's application to safety leadership.

The Goals and Objectives Principle

Meaningful goals and objectives can be powerful motivational forces. For maximum impact, both the goals and the objectives are necessary. Goals are *general*, objectives are *specific*. Goals relate to vision, mission, mental images of what can be, the overall results to be achieved. For example, here are six goals, related to the Environment. Health, and Safety (EHS) department's mission statement, for an international firm with about 65,000 employees:

- Use worldwide EHS standards to manage all facilities within a consistent framework.
- Ensure awareness and compliance with all applicable EHS laws, regulations and standards by means of ongoing communication and training programs.
- Incorporate both management information and a EHS data base into the information reporting system.
- Maintain an assessment system for all existing facilities and all acquisitions to assure compliance with laws, regulations, company policies and standards.
- Pursue a "zero waste" state through a measurable waste minimization process that results in efficient use of

This chapter contributed by George L. Germain. MA, CHCM, ASA, Executive Consultant for DNV-Loss Control Management, Loganville, GA.

Practical Principles for Managing Motivation

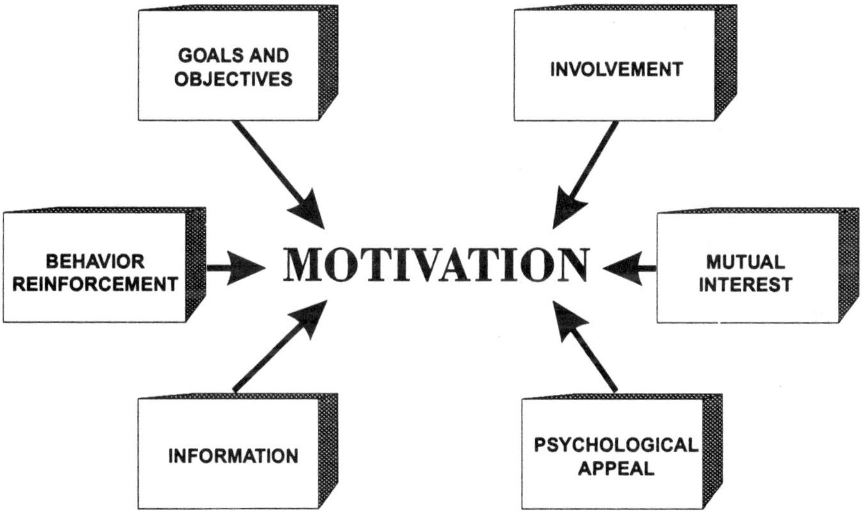

GOALS AND OBJECTIVES

INVOLVEMENT

BEHAVIOR REINFORCEMENT

MOTIVATION

MUTUAL INTEREST

INFORMATION

PSYCHOLOGICAL APPEAL

The Goals and Objectives Principle - Motivation to accomplish results tends to increase when people have meaningful goals toward which to work.

The Principle of Involvement - Meaningful involvement increases motivation and support.

The Principle of Mutual Interest - Programs, projects and ideas are best sold when they bridge the wants and desires of both parties.

The Psychological Appeal Principle - Communication that appeals to feelings and attitudes tends to be more motivational than that which appeals only to reason.

The Information Principle - Effective communication increases motivation.

The Principle of Behavior Reinforcement - Behavior with negative effects tends to decrease or stop; behavior with positive effects tends to continue or increase.

Figure 7-2

This chapter contributed by George L. Germain. MA, CHCM, ASA, Executive Consultant for DNV-Loss Control Management, Loganville, GA.

resources and minimum loss.

- Provide corporate EHS resource capabilities and consultation to help operational managers predict, identify, solve and prevent EHS problems.

To make EHS goals most meaningful to operational managers, tie them in with the company's strategic goals.

The power of goals is reflected in this quotation from *Managing For Excellence:*

> An overarching goal helps *keep the leader and members focused* on the *larger issues*. Managers can easily be swamped by the day-to-day minutiae of procedures, rules, deadlines, and other annoyances and lose sight of the department's reasons for existence, but continued reference to the goal guards against tunnel vision.... Further, the overarching goal is important for its *motivational properties*. When the departmental task is defined in terms of a challenge that has a larger meaning, involvement goes up. Most people need to believe in something that is larger than their day-to-day, often mundane, tasks. They are more likely to become committed to making things happen right if they believe in the significance of the unit's goal.

But goals are general. They need to be broken down into more detailed objectives. To help ensure maximum motivational values from objectives, make them **S–M–A–R–T:**

Specific - Not "We've got to improve our safety performance," but more like "Reduce our accidental damage losses by 20% this year."

Measurable - There must be standards and indicators

This chapter contributed by George L. Germain. MA, CHCM, ASA, Executive Consultant for DNV-Loss Control Management, Loganville, GA.

which show whether or not the standards are met. Whenever possible, these measures should be in terms of numerical, percentage, time or other finite figures, that is, quantifiable. Not "To improve the safety attitudes within the department," but more like "Ensure that supervisors/ team leaders give a safety tip every time they assign a critical task."

Attainable - When objectives are set unrealistically high (such as "Have zero accidents and incidents on our 47 construction sites this year"), they have little or no positive motivational value. People laugh at them, ignore them, or are demoralized by them. Motivational objectives are attainable objectives (Such as "Reduce our lost time accidents by at least 30% this fiscal period").

Relevant - Meaningful objectives are clearly related to the goals being pursued. People should be able to see how fulfilling the objectives will help meet the goals. Effective objectives should also be relevant to the individual, i.e., have personal meaning for the individual, be accepted by the individual, and be something for which attainment can be influenced by the individual.

Time-bounded - People are not positively motivated when objectives are too remote or indefinite. For example, the promise of a gold watch for 30 years of accident-free performance has no impact on the day-to-day efforts of workers. To motivate people, have objectives for this month, this week, this shift, and perhaps this hour. Use dates and deadlines.

People perform more effectively and enthusiastically when they have meaningful goals and measurable objectives to shoot for... either in sports or at work.

This chapter contributed by George L. Germain. MA, CHCM, ASA, Executive Consultant for DNV-Loss Control Management, Loganville, GA.

> To go from a zero
> To a management hero
> To answer these questions you'll strive:
> Where am I going?
> How will I get there?
> And, how will I know I've arrived?
>
> - Robert F. Mager

The Principle of Involvement

Meaningful involvement increases motivation, commitment and support. To stimulate involvement, ask people for suggestions, recommendations and advice in matters that affect their work. This helps to develop mutual interest, a climate of collaboration and cooperation. Such involvement is packed with motivational power. People tend to develop a feeling of ownership and support of what they helped to create. This power is evident in shift safety teams, loss control projects teams, quality circles, progressive safety committees, and other forms of participative problem solving teams.

Managers, team leaders and loss control professionals who use this principle effectively develop mutual interest, mutual motivation, and mutual respect. It pays big dividends to periodically inventory the level of involvement and ownership you develop in others. William Dyer refers to research evidence along these lines in *Contemporary Issues in Management and Organization Development*: A great deal of research has shown that a major way of changing the work environment is to increase the participation of people in planning, goal-setting, and decision-making processes. There has been clear research evidence that the motivational forces for many people are stimulated when they are involved in determining the cause of their own actions....

This chapter contributed by George L. Germain. MA, CHCM, ASA, Executive Consultant for DNV-Loss Control Management, Loganville, GA.

When there is honest involvement people being asked for their ideas and suggestions and these ideas being listened to and utilized—participation has been noted time after time to result in improved performance.

The Principle of Mutual Interest

Programs, projects and ideas are best sold when they bridge the wants and desires of both parties. People who are best at "selling" programs, projects or ideas are those who clearly establish a bridge or connection of values between what they want and what the other person(s) want. They focus on the benefits of the idea or program, especially for the other party, and build upon them, honestly and persistently. Everyday sayings that reflect this practical motivation principle include ones like these: "You help me, I'll help you." "You scratch my back and I'll scratch yours." "Here's how you can help the company and help yourself at the same time."

Leaders who want to make maximum use of this-principle could show their mutual concern by asking their work team members to provide the information requested on the *Mutual Aid Motivation* sheet (*Figure* 7-3).

The Psychological Appeal Principle

Communication that appeals to feelings and attitudes tends to be more motivational than that which appeals only to reason. Remember the C-A-B framework? The "A" stands for "Affective" — what most people call "emotional." It has to do with feelings, emotions, beliefs, values, convictions, mood, morale, disposition and attitudes. The affective aspects of motivational appeals are just as important as the cognitive, perhaps even more important.

Even if your message is factual or "scientific," people will

This chapter contributed by George L. Germain. MA, CHCM, ASA, Executive Consultant for DNV-Loss Control Management, Loganville, GA.

Mutual Aid Motivation

1. What do I do as your supervisor or team leader, and what does the company do, that helps you the most in your job?

2. What do I do as your supervisor or team leader, and what does the company do, that hinders you the most in your job?

3. What can you do that will help me, as your supervisor or team leader, do the best job for the company?

Figure 7-3

This chapter contributed by George L. Germain. MA, CHCM, ASA, Executive Consultant for DNV-Loss Control Management, Loganville, GA.

listen and understand better if you find an emotional peg upon which to hang that understanding. The way people feel about things strongly affects the way they think about them. A sincere emotional appeal can produce empathy, understanding, motivation, and action most quickly and effectively.

> Effective motivation
> works through "the heart"
> as well as the mind.

The Information Principle

Effective communication increases motivation. When people understand clearly the results they are trying to accomplish and how their efforts contribute to those results, motivation increases. In addition, the leader who makes a sincere effort to both listen to people and keep them informed is telling them, "I think you are important. I want to be sure both you and I know what's going on."

Loss control professionals take every opportunity to motivate executives and leaders through effective communication. Some have, for instance, been able to implement a standard saying that "Safety (and/or Health and Environment) will be a part of every regularly scheduled management meeting, at every organizational level." This often means, also, that the professional attends some of those meetings, thereby increasing "visibility" and learning more about management strategies, goals and objectives. This helps in making maximum use of The Principle of Mutual Interest, i.e., interests in common between the Loss Control department and the rest of the management team.

The Principle of Behavior Reinforcement

Behavior with negative effects tends to decrease or stop;

This chapter contributed by George L. Germain. MA, CHCM, ASA, Executive Consultant for DNV-Loss Control Management, Loganville, GA.

behavior with positive effects tends to continue or increase. Experimental findings and good sense both support the conclusion that behavior is influenced by its consequences. If the consequences of what you do are *negative* (displeasure — pain — penalty — punishment — frustrated desires, goals and objectives), you will tend *not* to repeat that behavior. The flip side is that if the consequences are *positive* (pleasure — reward — recognition satisfied desires, goals and objectives), you will tend to *repeat* that behavior.

Applying this directly to safety, for example, it may be helpful to think of safe behavior reinforcement as "the immediate behavioral recognition of safety performance." The words "immediate" and "behavioral" are especially important. For maximum effect, the recognition should immediately follow the behavior. "Behavioral" recognition differs from general, personal recognition ("You're a good man, Charlie Brown."). Behavioral recognition is related to something definite the person has done to certain actions, to specific behavior (("Charlie, I sure do appreciate your attention to eye safety practices... like the way you used those cup goggles today, on every job that called for them. Keep it up.") Experiments and experience have shown that this type of behavior reinforcement can be a practical, powerful motivational aid.

> Challenge: make it a habit to, at least once every day, purposely try to catch someone doing something for safety . . . and let them know you caught them.

Here are some suggestions for continuous improvement of behavior reinforcement activities:

- In connection with inspection and performance observation activities, give increased attention to safe, proper acts.

This chapter contributed by George L. Germain. MA, CHCM, ASA, Executive Consultant for DNV-Loss Control Management, Loganville, GA.

- Invest more time in recognizing safe, proper performance. Give at least as much emphasis to positive acts and conditions as is given to the negatives (unsafe acts and conditions).
- Record more commendations.
- Emphasize the personal, positive aspects and benefits of proper, safe performance.
- Emphasize ways in which proper, safe behavior also helps the work group, the department, and the organization as a whole.
- Give tangible rewards for safe, proper behavior.
- Give psychological rewards for safe, proper behavior.
- Increase positive attitudes and behaviors by increasing employee participation in safety planning, implementation, assessment and improvement.
- Make a habit of safety behavior reinforcement so that others will make a habit of safe behavior.

> Habits are first cobwebs, then cables. Build positive behavior habits by repetition - repetition - repetition.

When properly applied, safe behavior reinforcement provides a plentiful supply of potential benefits, such as these:

- Increases safe, proper acts and their resultant conditions.
- Reduces need for punishment and its side-effects.
- Improves job satisfaction — attitudes — morale.
- Reduces gripes, grievances and groaning.
- Boosts the image of leaders — supervisors — managers — "the system".
- Cuts accidents, harm to people, damage to property, and related losses.

This chapter contributed by George L. Germain. MA, CHCM, ASA, Executive Consultant for DNV-Loss Control Management, Loganville, GA.

- Stimulates pride of positive performance.

COMMITMENT - ATTITUDES - HABITS

Nearly everyone says "management commitment" is crucial to the success of their program or process, whether it be for safety, quality, production or cost control. But what do they mean by commitment? And how does one create and nurture it? In exploring answers to questions such as these, we will include several points from the *Commitment* book by Bird and Germain. Here, for instance is a functional definition:

> "Total willingness to persistently devote all required resources to accomplishment of a mission."

It ties in with current concepts such as *mission, visionary leadership* and *culture.*

Commitment is a personal value, a habitual frame of reference, an "attitude of mind" as phrased by one of the most renowned psychologists of all times, William James. In his words:

> "The biggest discovery of my generation is that human beings can alter their lives by altering their attitudes of mind.
>
> As you think, so shall you be." William James

This is an aspect of psychology that has been ignored or denied by the behavioral purists:

...the influence of attitudes on actions.
...the effects of beliefs on behavior.
...the power of mind over matter.

But good sense and personal experience leave little, if any, doubt about the vital importance of attitudes such as commitment to a cause (a cause such as safety). An attitude is a habit of thinking that leads to behaviors which fit that

This chapter contributed by George L. Germain. MA, CHCM, ASA, Executive Consultant for DNV-Loss Control Management, Loganville, GA.

thought pattern. It's a readiness to respond in a certain way, a "mental set." Attitudes serve as a frame of reference in determining a person's interpretation of words, actions, conditions and events. A person's perception is that person's reality. Since people react to situations as they perceive them, it's easy to see that attitudes influence actions, beliefs beget behaviors.

The Commitment Process

Figure 7-4 (from p. 34 of *Commitment*) depicts four phases of the commitment process. Let's take an example through these four phases: a scenario in which a safety professional is working to develop management's commitment to "accident prevention" rather than their current concept of "injury/illness prevention."

AWARENESS - In personal contacts and management meetings, the safety professional explores and explains an accident definition that emphasizes "damage" as well as "harm to people," such as... ple," such as...

> ... an event, that results in unintended harm or damage.

The discussions also emphasize the high cost of preventable property damage, in terms of replacement costs, repair costs, downtime and related losses. Communications also focus on the fact that most accidents which result in "damage" have high potential for "harm to people." So, for both monetary and humane reasons, *total* accident control should be the goal.

ACTIVATION - The loss control safety professional works with selected team leaders, supervisors, and/or middle managers to analyze purchasing and maintenance records for the cost of accidents in their areas. They use a guideline such as...

This chapter contributed by George L. Germain. MA, CHCM, ASA, Executive Consultant for DNV-Loss Control Management, Loganville, GA.

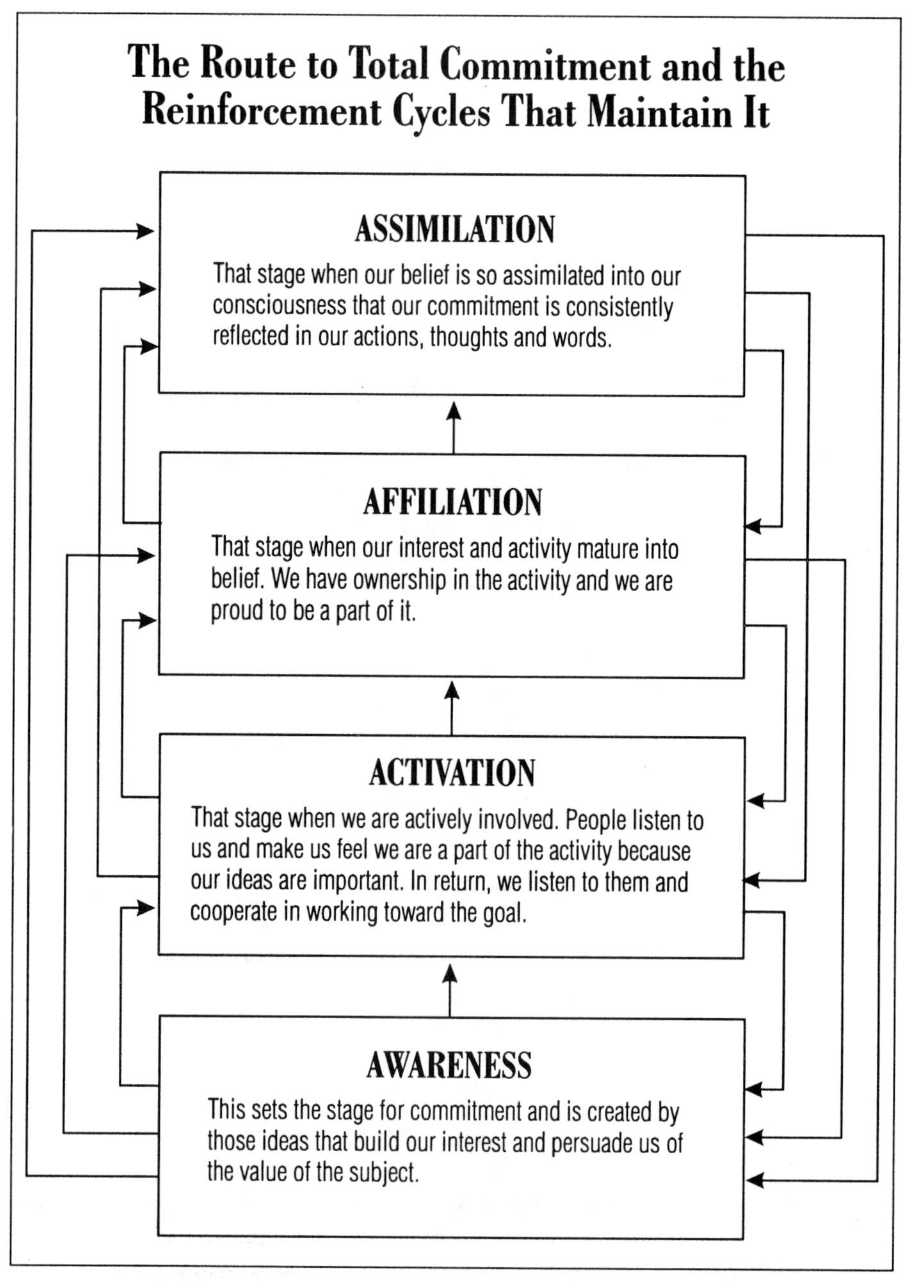

Figure 7-4

This chapter contributed by George L. Germain. MA, CHCM, ASA, Executive Consultant for DNV-Loss Control Management, Loganville, GA.

> ... any damage incident that is considered outside the standards established or desired for fair wear and tear by the most knowledgeable persons will be considered accidental, and will be included in the accident/incident reporting and investigation system.

This involvement and collaboration activates increased understanding and acceptance of the broader view sought by the safety professional, i.e., accident prevention rather than the limited concept of injury/illness prevention.

AFFILIATION - Based on their facts, figures and analyses, the initial interest of the collaborators grows into true belief that managing the system for total accident prevention is the professional path to pursue. These champions of the cause accept it as their own system and help to get it implemented throughout the organization.

ASSIMILATION - After a while, commitment to TAP (Total Accident Prevention) is so ingrained that it is automatic. It's simply "the way we do things around here." And the management team can move on to another challenge, another aspect of continuous safety improvement.

Do you have a passion for safety? Does your behavior demonstrate your devotion to control of accidental loss? Do you believe that the way to increase commitment from others is to share your commitment with them? As an unidentified sage put it, "attitudes are contagious... caught, not taught."

Figure 7-5 can be a good starting point for a reality check of your personal commitment. Notice that it includes many concepts critical to commitment: mission goals — objectives — affirmations — plans — beliefs — resources — expectations — work persistence. We suggest that you complete it, not "later" but right now! Clarify your commit-

This chapter contributed by George L. Germain. MA, CHCM, ASA, Executive Consultant for DNV-Loss Control Management, Loganville, GA.

PERSONAL COMMITMENT CHECK

	Yes	No
1. Do I have a clear mental picture of the mission?	______	______
2. Are my objectives specific?	______	______
3. Are my objectives measurable?	______	______
4. Are my objectives attainable?	______	______
5. Are my objectives relevant?	______	______
6. Are my objectives time-bounded?	______	______
7. Have I written specific affirmations to support my goals and objectives?	______	______
8. Do I repeat or renew my affirmations daily?	______	______
9. Do I have a definite plan for progress?	______	______
10. Do I use mental visualization (imaging to see what success will look like?)	______	______
11. Do I devote adequate actions to attainment of goals and objectives?	______	______
12. Do I really believe that the mission can be met?	______	______
13. Do I really believe that "I can...?"	______	______
14. Do I make the necessary sacrifices?	______	______
15. Do I devote adequate time, effort and resources to attain the end results I seek?	______	______
16. Does time seem to pass quickly when I'm working toward my goals and objectives?	______	______
17. Do I strengthen my desire and determination by working to overcome any barriers and setbacks?	______	______
18. Does my dedication and enthusiasm for the mission seem to be contagious?	______	______
19. Do I use the power of positive expectation?	______	______
20. Do I persist, persist, persist?	______	______
Total Score (5 points for each "yes")	______	______

Figure 7-5

ment strengths and considerations for improvements. You might also wish to use this figure (or your modification) as an aid to commitment checks by other leaders, supervisors and executives in your organization.

Six Ways To Influence Attitude Changes

In working with others to influence attitude changes, you can increase your impact by applying these helpful hints:

This chapter contributed by George L. Germain. MA, CHCM, ASA, Executive Consultant for DNV-Loss Control Management, Loganville, GA.

1. **Provide facts — knowledge — understanding.**

Effective communication, to increase awareness and understanding, is a major responsibility of loss control professionals. People need to know things such as:

- Definitions for terms like accident — incident — lost time injury — occupational illness environmental degradation — safety.
- Applicable laws, regulations, rules and standards.
- A loss causation model.
- Common causes of accidents and incidents (both immediate causes or symptoms and basic or root causes).
- Strategies and techniques for accident/incident investigation — facility inspection safety talks — orientation — effective task instruction — formal training — task analysis and procedures — planned performance observation and review — coaching problem solving — team building.
- Interpretation of Material Safety Data Sheets.
- Hazardous exposures and controls.
- Emergency preparedness.
- Housekeeping - cleanliness - order.
- The elements of safety management success.

As we saw earlier in The Information Principle, effective communication increases motivation.

2. **Change their role or perspective.**

You probably know of cases where the person's perspective changed considerably when they moved:

- from the job of equipment operator to the job of front line supervisor;
- from a union leadership position to a company

This chapter contributed by George L. Germain. MA, CHCM, ASA, Executive Consultant for DNV-Loss Control Management, Loganville, GA.

management position;
- from a staff job to a line job;
- from front line supervision to middle management, or middle management to top executive;
- from individual performer to true team member; or
- from team member to team leader.

> Different roles and responsibilities produce different perspectives and perceptions.

If you cannot actually change the person's role, you can use "empathy" to affect their attitude. First, you do your best to empathize with that person, to see things through their eyes, from their perspective.

> "Seek first to understand, then to be understood."
> - Steven R. Covey

Then you can help to develop their empathy for those who serve other roles. To help do this, you can use tools such as: communication – education – committee assignments – role playing – "assistant to" – positions mentoring – task forces problem solving teams.

3. Increase their involvement.

Four of the most motivational words in the world are, "I need your help." Loss control professionals can use those words to increase commitment, motivation, activities and results. Get workers, team leaders, supervisors and executives involved in safety program planning, investigation, root cause analysis, inspection, task analysis, safety meetings, emergency preparedness, safety promotion, training needs analysis, training sessions, accident imaging, incident

This chapter contributed by George L. Germain. MA, CHCM, ASA, Executive Consultant for DNV-Loss Control Management, Loganville, GA.

recall, and the whole process of risk identification, evaluation and control.

In other words, apply The Principle of Involvement: "meaningful involvement increases motivation, commitment, and support." We could just as well say "involvement increases commitment."

4. **Attend to the affective aspects.**

There used to be a widespread impression that business and industrial leaders deal primarily with facts, figures and logic... the cognitive or rational aspects of life. Nothing could be farther from the truth! Emotion is every bit as important as reason, at work as well as anywhere else. Here are some examples showing how various respected authors have highlighted the criticality of affective aspects (from William D. Hitt, Thoughts on Leadership (A Treasury of Quotations):

> "The human being needs a framework of values, a philosophy of life, a religion or religion-to live by and understand by, in about the same sense that he or she needs sunlight, calcium, or love. "
>
> - *Abraham Maslow*
> *Toward a Psychology of Being*

> "Let us suppose that we were asked for one all-purpose bit of advice for management, one truth that we were able distill from the excellent companies research. We might be tempted to reply, "Figure out your value system. Decide what your company stands for."...Clarifying the value system and breathing life into it are the greatest contributions a leader can make."
>
> - Thomas Peters and Robert Waterman
> *In Search of Excellence*

> "What can managers do that will show another

This chapter contributed by George L. Germain. MA, CHCM, ASA, Executive Consultant for DNV-Loss Control Management, Loganville, GA.

> person that the manager is truly concerned about him or her as a person? (1) talk with people — which means asking, listening, and sharing; (2) trust people with significant work; (3) allow people to influence their work world; (4) reward people for good work; (5) help people plan a future; and (6) take time to 'coach' people. "
>
> - William Dyer
>
> *Contemporary Issues in Management and Organization Development*

Today's business books and articles are filled with concepts such as vision, values, care, concern, compassion, trust, ethics, and self-esteem. As an example, the Blitzer, Petersen and Rogers article, "How To Build Self-Esteem," focuses on how to make people feel: (1) uniquely valuable, (2) competent, (3) secure, (4) empowered, and (5) connected to the group. Again, a definite concentration on affective aspects.

Figure 7-6 shows what Mildred Ramsey (author of *The Super Supervisor*) calls the "7 C's of Inside-out Motivation." Notice the emphasis on "matters of the heart." A key point is to remember and use The Psychological Appeal Principle:

Communication that appeals to feelings and attitudes tends to be more motivational than that which appeals only to reason.

5. **Model the behaviors that demonstrate desired attitudes.**

In *Practical Loss Control Leadership*, Bird and Germain include 12 fundamental truths or proven principles of leadership. One of them is "The Principle of Leadership Example":

> **People tend to emulate their leaders.** Most people want to please their leaders, and do so by

This chapter contributed by George L. Germain. MA, CHCM, ASA, Executive Consultant for DNV-Loss Control Management, Loganville, GA.

The Super Supervisor's 7 C's of Inside-Out Motivation

Courtesy - Employees expect a friendly, enthusiastic greeting every day. *Enthusiasm is caught . . . not taught. Start an epidemic.*

Concern - Employees have three dimensions: Body (what they are doing); Mind (what they are thinking); and Spirit (what they are feeling). For peak performance, a supervisor must relate to all three. *Employees are persons . . . not personnel.*

Consideration - Employees' feelings must be considered. Encourage them to express their views on all matters that concern them. *Encouragement energizes, discouragement paralyzes.*

Compassion - Employees need sympathy and compassion. Along with their families, they will inevitably experience the pain of accidents, illness and death. *Compassion is a powerful motivator that pays big dividends in performance.*

Consistency - Employees respond to a supervisor who uses the same tone of voice with everyone, treats everyone as an equal and is always pleasant and peaceful. *Harmony oils the wheels of industry - friction puts sand in the gearbox.*

Control - Employees work best for a supervisor who is confident and in control. Good supervisors control their tempers (no angry outbursts), their tongues (no gossiping) and their hormones (no flirting). A supervisor must control himself or herself before controlling others. *A reputation is made from many acts, but can be lost by just one.*

Caring - Employees respect supervisors who have made "caring" a lifestyle. A good supervisor cares about herself or himself (positive self-image), acts like a winner (self-confident) and plays by the rules (self respect). *Caring about employees and their families is a full-time job, week in and week out, day after day.*

Figure 7-6

This chapter contributed by George L. Germain. MA, CHCM, ASA, Executive Consultant for DNV-Loss Control Management, Loganville, GA.

> following their behavioral example. Attitudes and influence, like waterfalls, flow downward. At all levels of management, the attitudes and actions of leaders is one the most powerful motivational forces in the world. (p. 46) The old cop-out, "Don't do as I do, do as I say" did not work in leading children toward adulthood — and it does not work in leading employees and team members toward desired performance.

People are influenced by what managers say about safety and, more importantly, what managers do for safety. *Figure* 7-7 shows 50 things managers can do to show their commitment. As a checklist, it has been used very effectively in communication/education sessions for upper managers and loss control professionals.

6. **Apply behavior reinforcement.**

Behavior with negative consequences tends to decrease or stop. So do what you can to...

> ... increase the negative consequences of behaviors which reflect undesirable attitudes; and
> ... reduce the positive consequences of behaviors which reflect undesirable attitudes.

Behavior with positive consequences tends to continue or increase. So do what you can to...

> ... increase the positive consequences of behaviors which reflect desirable attitudes; and
> ... reduce the negative consequences of behaviors which reflect desirable attitudes.

To put it another way, make appropriate use of the following four types of consequences:

1. Positive reinforcement — increases behavior by providing pleasurable consequences (the person gets

This chapter contributed by George L. Germain. MA, CHCM, ASA, Executive Consultant for DNV-Loss Control Management, Loganville, GA.

MANAGEMENT SAFETY ACTIVITY CHECKLIST
(Actions Reflect Attitudes)

D. Not necessary for us
C. Have not done this, but should
B. Done, but should be improved
A. A well-met requirement

Administrative Actions	**A**	**B**	**C**	**D**
1. Issue a policy statement reflecting management's positive attitude and commitment to loss control.	___	___	___	___
2. Appoint specific persons as loss control coordinators.	___	___	___	___
3. Establish joint safety & health committee and/or safety & health representatives.	___	___	___	___
4. Establish an internal written procedure for receiving, considering, and following-up recommendations from the joint committee and/or representatives.	___	___	___	___
5. Establish an internal written procedure for handling refusal to work situations.	___	___	___	___
6. Issue loss control manual to guide all levels of management on matters of program policy, procedures and practices.	___	___	___	___
7. Incorporate specific responsibilities into all management job descriptions.	___	___	___	___
8. Establish specific, measurable standards of performance in the loss control program.	___	___	___	___
9. Establish annual loss control goals and objectives for the organization.	___	___	___	___
10. Establish and implement an internal written standard requiring loss control orientation for all management personnel.	___	___	___	___
11. Require and support training (both initial and review) in loss control for all management personnel.	___	___	___	___
12. Ensure adequate training, by a recognized outside agency, loss control coordinators.	___	___	___	___
13. Establish an internal written procedure for management review of inspection reports and follow-up of remedial actions.	___	___	___	___

This chapter contributed by George L. Germain. MA, CHCM, ASA, Executive Consultant for DNV-Loss Control Management, Loganville, GA.

	Administrative Actions	**A**	**B**	**C**	**D**
14.	Issue a written directive requiring task analyses and the development of work procedures and practices for all critical tasks.	___	___	___	___
15.	Issue a directive which requires a task observation program and stresses its importance for accident control.	___	___	___	___
16.	In writing, designate a coordinator to administer the overall emergency plan.	___	___	___	___
17.	Ensure that departmental or sectional coordinators are appointed to administer the emergency plan within their areas of the organization.	___	___	___	___
18.	Ensure a comprehensive written emergency plan that is kept current with changed conditions, facilities, equipment, processes and materials.	___	___	___	___
19.	Establish a written policy covering discipline for non-compliance with rules, and commendation for good compliance.	___	___	___	___
20.	Establish a policy or practice governing the proper use and conservation of personal protective equipment.	___	___	___	___
21.	Ensure that written procedures and practices and special instructions are given to appropriate managers and employees exposed to potential health hazards.	___	___	___	___
22.	Ensure that the company's purchasing policy, or senior management directive, includes loss control considerations.	___	___	___	___
23.	Establish an engineering policy or executive directive requiring the loss control coordinator's review for safety and health considerations at the conception and design stages of all new development and construction or modifications.	___	___	___	___
24.	Establish a system to analyze the impact on the organization of off-the-job accidents.	___	___	___	___
	Personal Involvement Actions				
25.	Refer to loss control policy in management meetings, training programs, written materials and personal contacts.	___	___	___	___
26.	Distribute loss control letters or memos to all employees at least quarterly.	___	___	___	___

This chapter contributed by George L. Germain. MA, CHCM, ASA, Executive Consultant for DNV-Loss Control Management, Loganville, GA.

	Personal Involvement Actions	**A**	**B**	**C**	**D**
27.	Make loss control tours at least once every six months.	___	___	___	___
28.	Attend safety meetings at least once every six months.	___	___	___	___
29.	Take part in at least annual audits of the loss control program.	___	___	___	___
30.	Take part, at the scene, in the investigation of all fatalities, major loss accidents and high-potential loss incidents.	___	___	___	___
31.	Take part in training drills for emergencies.	___	___	___	___
32.	Set the leadership example regarding rules compliance, use of personal protective equipment, and adherence to safe work practices.	___	___	___	___
33.	Appoint problem solving teams to solve critical problems of property damage, as well as occupational injury and illness.	___	___	___	___
34.	Maintain "visibility" in occupational health activities, through written communications, special tours, attendance at company meetings on the topic, and maintenance of high occupational health standards.	___	___	___	___
35.	Make special presentations to various employee groups on specific loss control topics at least twice a year.	___	___	___	___
36.	Chair periodic safety promotion theme campaigns.	___	___	___	___
37.	Present safety awards and recognitions.	___	___	___	___
38.	Personally promote on-the-job training for off-the-job safety leadership.	___	___	___	___
	Two-Way Feedback Actions				
39.	Discuss with subordinates their program elements and activities for implementing the loss control policy.	___	___	___	___
40.	Discuss procedures, practices, problems and progress with loss control coordinators.	___	___	___	___
41.	From each organizational head, require development and periodic updating of a thorough inventory and evaluation of loss exposures.	___	___	___	___
42.	Require and discuss with subordinates the loss control objectives for their organization components.	___	___	___	___

This chapter contributed by George L. Germain. MA, CHCM, ASA, Executive Consultant for DNV-Loss Control Management, Loganville, GA.

	Two-Way Feedback Actions	**A**	**B**	**C**	**D**
43.	In performance reviews, salary administration activities and career development discussions give as much weight to loss control performance as is given to performance for quality, production and cost control.	___	___	___	___
44.	Include loss control on the agenda of all regularly scheduled management meetings.	___	___	___	___
45.	Conduct review meetings of all major investigations and communicate findings to other management personnel.	___	___	___	___
46.	Follow-up and discuss the activities and results of appointed loss control problem solving teams.	___	___	___	___
47.	For evaluating and discussing loss control performance, require and use not only measurements of consequence but also measurements of cause and measurements of control.	___	___	___	___
48.	Use measurement data for performance feedback and coaching.	___	___	___	___
49.	Apply enforcement (constructive correction) and reinforcement (commendation) techniques to maintain compliance with performance standards.	___	___	___	___
50.	Review and act upon the findings of periodic audits of the management system for loss control.	___	___	___	___
	Additional Actions				
51.					
52.					
53.					
54.					
55.					

Figure 7-7

This chapter contributed by George L. Germain. MA, CHCM, ASA, Executive Consultant for DNV-Loss Control Management, Loganville, GA.

something he/she wants).
2. Negative reinforcement — increases behavior by reducing unpleasant consequences (the person avoids something he/she does not want).
3. Punishment — decreases behavior by providing unpleasant consequences (the person gets something he/she does not want).
4. Extinction — decreases behavior by failing to provide pleasurable consequences (the person does not get anything he/she wants).

FIVE KEYS TO MOTIVATION MANAGEMENT

Key #1 - Clarify and Communicate Expectations

> Good leaders know that if they expect a lot of their constituents, they increase the likelihood of high performance. That means standards, an explicit regard for excellence. In a high-morale society, people expect a lot of one another, hold one another to high standards. And leaders play a special role in conveying such expectations.
>
> Good leaders don't ask more than their constituents can give, but they often ask — and get — more than their constituents intended to give or thought it was possible to give.... High performance takes place in a framework of expectation.
>
> —John Gardner

What do executives expect of the organization? Of their management team members? Of loss control professionals? Of non-exempt employees? What do loss control professionals expect of supervisors and executives? Of other employees? What do employees expect of the organization? Of their team leaders - supervisors - upper managers? Of loss control professionals? These expectations need to be

This chapter contributed by George L. Germain. MA, CHCM, ASA, Executive Consultant for DNV-Loss Control Management, Loganville, GA.

clarified and communicated.

Expectations are a vital piece of the motivational puzzle. They relate to everything from the overarching vision or mission, through group and individual performance standards, to specific job and task responsibilities.

Here is an excerpt from the Corporate Loss Prevention Manual of a large chemical company, setting the stage for expectations throughout the organization:

> The first premise is that safety is an integrated function of line management. It is not an adjunct to management, the responsibility for which can be delegated to staff persons or employees. The safety department is committed with line managers for safety results and will provide advice and assistance. Full cooperation from each employee is necessary and expected. However, line managers are accountable for their decisions, actions and results. Every manager in the company is responsible both for the safety of his or her people, and the safeness of all facilities or work places which the company has placed under his or her control. This philosophy is depicted in the [Company] Safety Logo which symbolizes that our people, around the world, must safety interact with people, equipment, materials, products and environment to accomplish our safety goals. The management of safety is little different from any other aspect of good management. In fact, our safety performance is one indicator of how well we are running our total business.
>
> The second premise is that safety must be a matter of management CONVICTION. It is not enough to be just "interested in" or "concerned

This chapter contributed by George L. Germain. MA, CHCM, ASA, Executive Consultant for DNV-Loss Control Management, Loganville, GA.

> about" safety. Rather, safety shall be regarded as equal to all other considerations. We firmly stand by the principle that, "No management function has any higher priority than the safety of people."

The manual then outlines key expectations (that is, responsibilities/accountabilities) for various management positions: Group President - Division Vice President, General Manager - Facility Manager - Production or Maintenance Superintendent - Front-line Supervisor - Safety Coordinator. One example (Facility Manager) is shown as *Figure* 7- *8.*

What should the company, the managers, the leaders and all employees expect of loss control professionals? Based on our work with hundreds of organizations, here are some suggestions regarding the role of those with functional responsibility for safety, health and/or environmental protection. The fulfillment of functions such as these motivates safety success.

1. Serve as an internal consultant. Build this role around these four A's... Analyze — Advise — Assist — Audit:
   - Analyze the organization's pertinent policies, procedures, practices, problems and progress.
   - Advise line management regarding loss control regulations, standards and requirements; as well as tools and techniques for effective control of accidental loss.
   - Assist line management in ensuring the communication, training and motivation that enable fulfillment of goals, objectives and responsibilities.
   - Audit how well the management system is working for/loss control, and help line managers develop action plans for improving that system.
2. Tie loss control goals and objectives in with operating management's vision, mission, strategies, goals and

This chapter contributed by George L. Germain. MA, CHCM, ASA, Executive Consultant for DNV-Loss Control Management, Loganville, GA.

FACILITY MANAGER

GENERAL RESPONSIBILITY:

The responsibility for the facility's safety performance is vested in the Facility Manager who is accountable to an executive officer. The Facility Manager:

1. Establishes and administers the Corporate Safety Program in the facility in keeping with predetermined objectives.
2. Establishes controls to assure uniform department performance in compliance with safety program elements.
3. Establishes a training program which will develop in each member of management a strong safety attitude and a clear-cut understanding of specific duties and responsibilities.

SPECIFIC ACTIVITIES:

1. Publishes a plant safety policy patterned after the corporate policy.
2. Issues general plant safety rules and regulations over his signature.
3. Reviews daily the plant safety performance' for the past 24 hours and discusses problems with staff.
4. Sees that compliance with established safety rules, regulations, procedures and practices is maintained.
5. Coordinates emergency activities in the event of a major incident.
6. Adopts a plant-wide housekeeping inspection program and defines geographic areas within the facility location which delineate responsibility for housekeeping.
7. Participates in the housekeeping program by monthly inspection of an assigned area.
8. Determines annually the facility's safety program effectiveness and the need for change to improve performance; establishes a 12-month Action Plan and assigns help as needed.
9. Includes safety as an agenda item in staff meetings.
10. Serves as chairman of the committee investigating lost time injuries and major disasters.
11. Establishes goals with Maintenance/Production Superintendents for the coverage of critical jobs with procedures and practices, and communicates these objectives.
12. Establishes a facility-wide emergency plan.
13. Establishes and defines a hazard committee, and requires their review of process equipment prior to construction and placing into operation.
14. Establishes quarterly and yearly assessments to measure and evaluate safety program compliance at the department and facility level.
15. Reviews and critiques monthly records of the plant's safety activities.
16. Makes management performance in the safety program a part of job descriptions and a factor in job appraisals.

Figure 7-8

This chapter contributed by George L. Germain. MA, CHCM, ASA, Executive Consultant for DNV-Loss Control Management, Loganville, GA.

objectives. Stay on their wavelength.

3. Stimulate active involvement in loss control efforts by all people, at all levels, in all departments.
4. Pursue continuous improvement of technical loss control expertise through networking, reading, training courses, membership and participation in professional organizations.
5. Develop and maintain a working knowledge of the company's organizational structure, facilities, equipment, materials, processes, products/services, and customers.
6. Show adequate concern for economics, as line managers must. For example:
   - Operate on a budget basis.
   - Use cost-benefit analyses for proposals, projects and programs.
   - Calculate and communicate the cost effectiveness of loss control efforts.
7. Collaborate with line management on:
   - Identifying the work required for loss control success, program elements — roles — responsibilities — accountabilities.
   - Adding internal standards/requirements to those required by law.
   - Using regular, representative measurements and periodic, comprehensive audits to assess the effectiveness of the management system.
   - Evaluating measurements and results to identify system strengths and performance improvement potentials.
   - Ensuring effective communication, coaching, constructive correction, and commendation regarding loss control performance.

This chapter contributed by George L. Germain. MA, CHCM, ASA, Executive Consultant for DNV-Loss Control Management, Loganville, GA.

8. Persistently pursue development and maintenance of a "safety culture" (or loss control culture) throughout the organization.

When we clarify and communicate expectations, we take a giant step toward harnessing the motivational power of positive expectation. Expectations become conscious goals. The expectation of specific behaviors helps to produce those behaviors. People tend to become what their leaders expect them to become. Low expectations tend to produce low performance. High expectations tend to produce high performance.

> Be careful about what you expect...
> you are likely to get it.

To motivate high performance levels, set high standards, communicate a contagious enthusiasm for meeting those standards, and demonstrate to others your belief in their ability to meet or beat the standards.

Key #2 - Provide Proper Resources and Support

Though *expectations* are powerful and important, they are only part of the puzzle. Managers must do more than say "Safety is as important as quality or production".

> Your actions shout so loudly
> that
> I can't hear what your words are saying!

Credibility is enhanced when leaders' deeds fit their words. Their true belief is reflected in their behavior.

Like any other worthy cause, loss control activities require resources. For optimum results, the management system must provide optimum resources, such as:

This chapter contributed by George L. Germain. MA, CHCM, ASA, Executive Consultant for DNV-Loss Control Management, Loganville, GA.

... visible involvement of managers/leaders

... enabling policies, procedures and practices

... an organizational structure which enables and encourages teamwork, trust, cooperation and collaboration

... administrative and staff support

... adequate numbers of people with the requisite knowledge and skills

... sufficient time, space, facilities, equipment, materials and money

... appropriate training and development

... delegation and empowerment

... a fair and equitable system for recognition and reward.

Without adequate resources and support, the probability of success is minimal. With adequate resources and support, the probability of success is maximal.

Key #3 - Emphasize Education, Training and Development

> The innovative organization requires a learning atmosphere throughout the entire business. It creates and maintains continuous learning. No one is allowed to consider himself "finished" at any time. Learning is a continuing process for all members of the organization.
>
> —Peter Drucker,
> *Management*

Referring back to *Figure 7-1*, page 209, reminds us that holistic motivation involves three interrelated domains:

This chapter contributed by George L. Germain. MA, CHCM, ASA, Executive Consultant for DNV-Loss Control Management, Loganville, GA.

ATTITUDES	KNOWLEDGE	SKILLS
• Feelings	• Information	• Words
• Emotions	• Ideas	• Actions
• Beliefs	• Thoughts	• Behavior
• Values	• Concepts	• Capabilities
• Convictions	• Mental pictures	• Performance

So far, we have given major emphasis to the "attitudes" domain. Now let's emphasize the "knowledge" and "skills" domains. It is amazing how many managers seem to believe in magic and miracles for continuous improvement of knowledge and skills . . . especially regarding front line supervisors and members of teams. They change a person's job from equipment operator on Friday to supervisor the following Monday. By magic and miracles, the person now is expected to exhibit all sorts of necessary skills, such as:

... conduct effective accident investigations

... give great safety talks

... conduct group meetings

... perform professional facility inspections

... prepare critical task analyses and procedures

... facilitate effective team results.

Or, several people are appointed to a team project. Through miracles and magic, they are expected to acquire knowledge and skills regarding:

... team roles and responsibilities

... facilitation skills

... how to handle conflict constructively

... problem-solving

This chapter contributed by George L. Germain. MA, CHCM, ASA, Executive Consultant for DNV-Loss Control Management, Loganville, GA.

... group dynamics

... how to analyze data and present it pictorially

... persuasive presentation of proposals and projects

... how to track and communicate results and benefits.

The knowledge and skills required to do those things that do not materialize magically or miraculously. They require continuous education, training and development. This has become so important that there are many speeches, seminars, books and videos predicting doom for the company that fails to become a *learning organization.*

The International Safety Rating System™ (ISRS®), has been used in thousands of organizations around the world to assess their systems. Of its 20 elements, two whole elements are devoted to this domain, one for "Knowledge and Skill Training" and one for "Leadership Training." Here are selected samples of the types of questions that must be answered and documented in ISRS® assessments:

- Has a systematic approach been taken for training needs analysis, including review of:

  ... work activities and responsibilities for all occupations?

  ... critical task analyses and procedures?

  ... hazard analyses?

  ... inspection report analyses?

  ... accident/incident analyses?

  ... task observation reports?

  ... change management analyses?

  ... operating procedures?

  ... applicable regulations, codes and standards?

  ... training program evaluation reports?

  ... surveys of employees and managers?

This chapter contributed by George L. Germain. MA, CHCM, ASA, Executive Consultant for DNV-Loss Control Management, Loganville, GA.

- How often are training needs formally reviewed and updated...

  ... for each occupation?

  ... for each individual?
- Have those who conduct training received trainer training?
- To what extent are written materials, audio-visuals, and other effective aids used in training programs?
- To what extent have lesson plans been developed to guide training content and quality?
- Are knowledge and proficiency tests used effectively in training programs?
- Does loss control training include topics such as...

  ... management control of loss?

  ... the causes and effects of loss?

  ... planned inspections and maintenance?

  ... critical task analysis and procedures?

  ... accident/incident investigations?

  ... task observation?

  ... rules and work permits?

  ... measurement techniques in loss control?

  ... knowledge and skill training?

  ... personal protective equipment?

  ... introduction to health and hygiene control?

  ... purchasing and transport controls?

  ... personal communications?

  ... group communications?

  ... property damage and waste control?

  ... environmental management?

This chapter contributed by George L. Germain. MA, CHCM, ASA, Executive Consultant for DNV-Loss Control Management, Loganville, GA.

... off-the-job safety?

- Are records maintained to show who has received...

  ... loss control orientation?

  ... loss control initial training?

  ... loss control refresher training?

- Are there formal refresher training programs which review critical segments of the training at least every three years?
- How well are training programs evaluated in terms of...

  ... participant reactions to the training?

  ... measured learning?

  ... on-the-job application?

  ... results (safety - quality - production - profitability)?

- Are the evaluation results used effectively for continuous improvement of training activities?

Training without development isn't worth much. By development we mean activities like practice, application, feedback, modeling, mentoring, coaching, facilitating and reinforcing.

The most powerful motivation for applying the knowledge and skills learned and/or honed in training programs is provided by the work environment — the management system. Learners need (as shown in *Figure 7-9*) O-S-C-A-R: opportunity, stimulation, counsel, assistance and reinforcement. The system should provide bridges to the new behaviors, not barriers. Here are some of the specific things that leaders/supervisors/managers can do to build those bridges, to stimulate application of training (and the resultant return on investment):

This chapter contributed by George L. Germain. MA, CHCM, ASA, Executive Consultant for DNV-Loss Control Management, Loganville, GA.

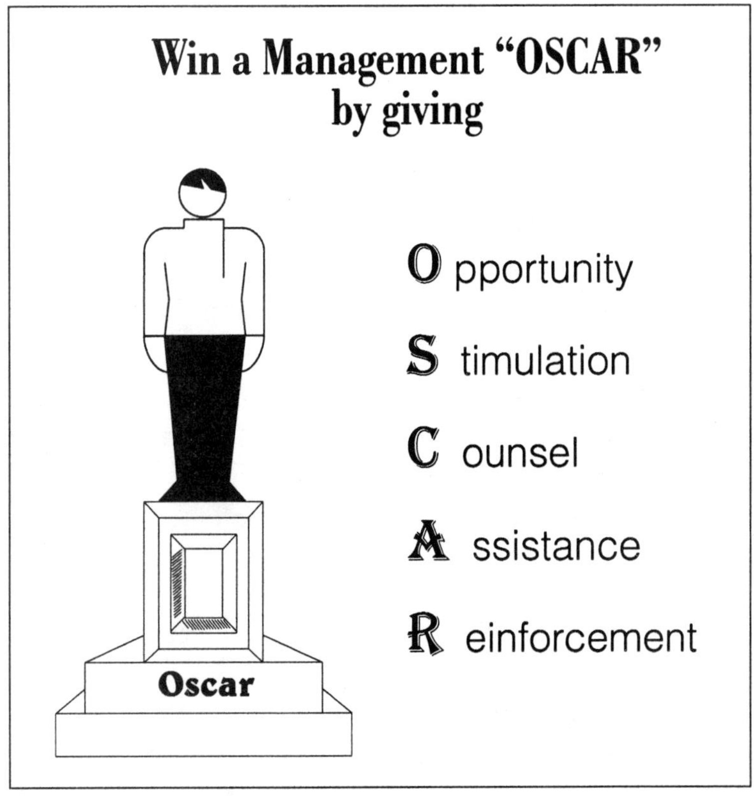

Figure 7-9

Before The Training

- Participate in the training needs analysis.
- Work with trainers and potential participants on program planning and development.
- Ensure that the philosophy of the training is consistent with that of the management system.
- Discuss with potential participants the reasons for the training, and the anticipated benefits to both the participant and the organization.
- Review the program objectives, content, format and

This chapter contributed by George L. Germain. MA, CHCM, ASA, Executive Consultant for DNV-Loss Control Management, Loganville, GA.

process. If possible, take part in an appreciation or overview session.

- If there are precourse assignments, allow time for program participants to complete them.
- Facilitate discussion between potential and past participants in the training program.
- Ensure a positive training environment (timing - location - facilities - equipment - amenities).
- Arrange work assignments to avoid a heavy backlog of work for returning program participants.
- Plan to participate, if feasible, in the training process.

During The Training

- Whenever possible take part in at least the program introduction and closing.
- Ensure minimal interruptions for the program participants.
- Demonstrate management interest and support.
- Plan for transfer of training from the training place to the workplace.
- Ensure that substitutes perform the critical activities to prevent an oppressive backlog for the training program participants.
- Monitor attendance, progress and reactions.

-After The Training-

- Have a debriefing session with the trainer.
- Discuss with participants what they learned.
- Have participants share key aspects of the training with team members and peers.
- Mutually develop and implement plans for applying the new knowledge and skills.

This chapter contributed by George L. Germain. MA, CHCM, ASA, Executive Consultant for DNV-Loss Control Management, Loganville, GA.

- Show tolerance and empathy for "less-than-perfect" performance and results. Recognize that skills development is a process, not an event. It takes time, feedback and practice.
- Model the desired behaviors.
- Coach for the desired behaviors.
- Recognize and reinforce the desired behaviors.
- Celebrate success!

Key #4 - Practice Performance Management

To motivate continuous improvement of performance and results, team leaders, supervisors and upper managers must continuously practice performance management. *Figure 7-10* summarizes the following eight phases of a performance management model for achieving continuous performance improvement.

- Phase 1: Present Situation Appraisal -

Baseline information is important. This is true whether you are concerned with managing the performance of the whole organization, a team within the organization, or one person. Processes that are helpful in obtaining this information include such widely-used tools as management system audits or assessments, team evaluations, and individual performance appraisals. The results show "what is," the present picture, the strengths to build on, and the performance improvement potentials. This baseline information also enables future comparisons to show the degree of change and improvement.

- Phase 2: Dream - Vision - Mission -

This involves painting a picture of the "desired situation"... for the organization, for the team, for the individual. This is an ideal and unique image of the future. Commitment to

This chapter contributed by George L. Germain. MA, CHCM, ASA, Executive Consultant for DNV-Loss Control Management, Loganville, GA.

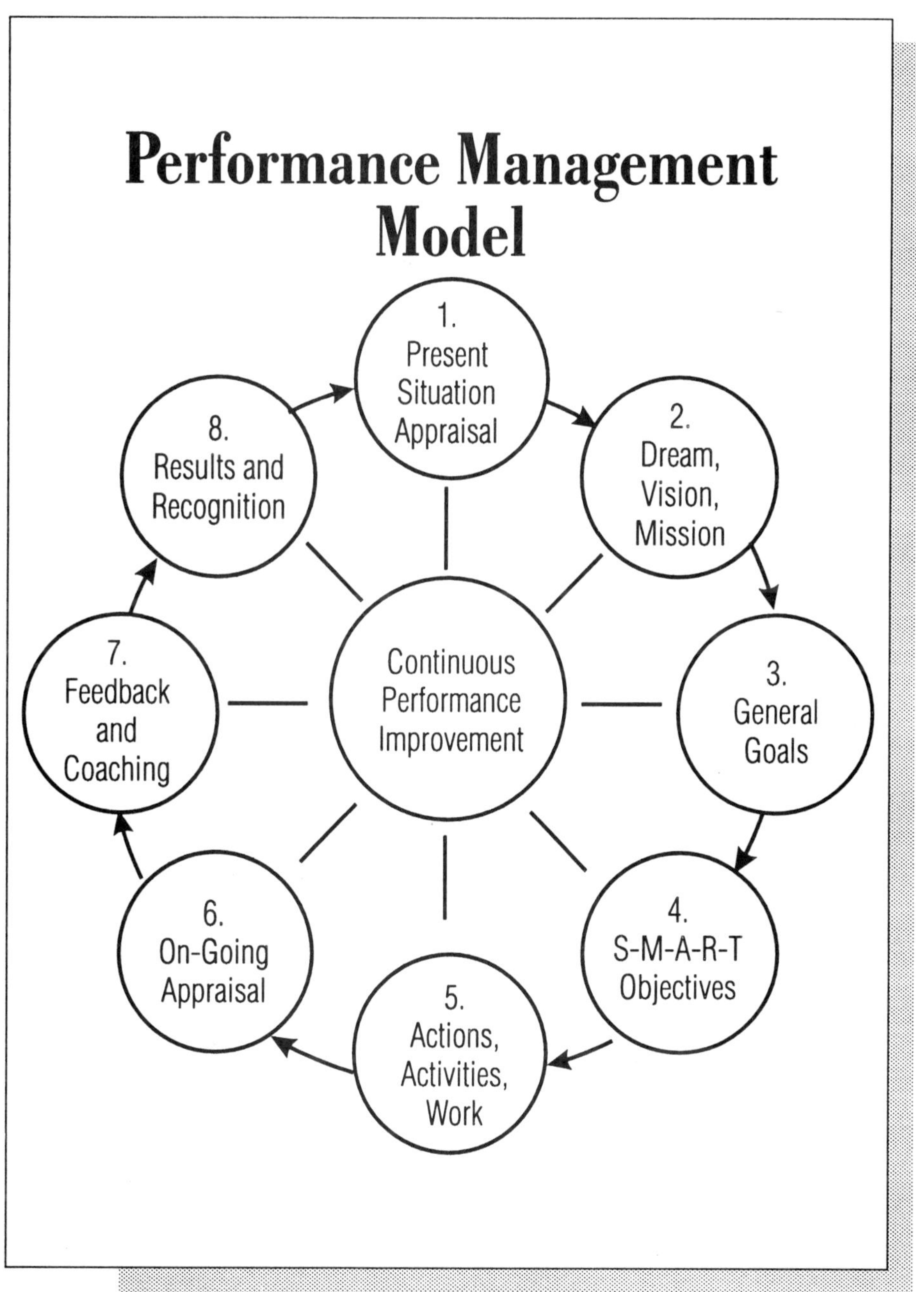

Figure 7-10

This chapter contributed by George L. Germain. MA, CHCM, ASA, Executive Consultant for DNV-Loss Control Management, Loganville, GA.

this mental picture is a powerful motivator, giving direction and purpose to the enterprise and a focus for positive performance.

Here is how Bennis and Nanus emphasized this phase (in *Leaders: The Strategies for Taking Charge)*:

> To choose a direction, a leader must first have developed a mental image of a possible and desirable future state of the organization. This image, which we call a vision, may be as vague as a dream or as precise as a goal or mission statement. The critical point is that a vision articulates a view of a realistic, credible, attractive future for the organization, a condition that is better in some important way than what now exists....
>
> If there is a spark of genius in the leadership function at all, it must lie in this transcending ability, a kind of magic, to assemble — out of all the variety of images, signals, forecasts, and alternatives — a clearly articulated vision of the future that is at once simple, easily understood, clearly desirable, and energizing.

Think of a vision as a lighthouse. Its message is, "As you journey to where you are going, keep me in view or you risk getting lost."

- Phase 3: General Goals -

This means to crystallize the generic results required to fulfill the dream, the vision, the mission. Say, for example, a company's loss control mission is to build and maintain a worldwide management system equal to or better than the benchmark systems within the industry. Their many goals might include ones such as these few examples:

- To develop and maintain a management system

This chapter contributed by George L. Germain. MA, CHCM, ASA, Executive Consultant for DNV-Loss Control Management, Loganville, GA.

which integrates SQPC (Safety - Quality - Production - Cost) concepts.

- To ensure awareness and compliance with all applicable laws, regulations and standards by means of ongoing communication and training programs.
- To incorporate a loss control data base into the management information system.
- To provide appropriate professional expertise and assistance to help line managers predict, identify, solve and prevent loss control problems.

- Phase 4: S-M-A-R-T Objectives -

General goals should be broken down into more detailed objectives. As discussed earlier in this chapter, objectives with maximum motivational power are specific, measurable, attainable, relevant and time-bounded. Example: "All supervisors and team leaders will be trained in root cause analysis before December 31 of this year."

- Phase 5: Actions - Activities - Work -

Mission, goals and objectives are worthless if they are nothing more than good intentions. As Peter Drucker put it, "Unless objectives are converted into action, they are not objectives, they are dreams." The purpose of determining the business mission, goals and objectives is not knowledge, but action... goal-directed, results-oriented work activities. To focus the organization's resources and energies on results, the work program must include concrete work assignments with deadlines, and with accountability.

Accountability is a critical key to effective motivation. Among other things, it requires:

1. Clear goals and objectives.
2. Agreed upon responsibilities and performance re-

This chapter contributed by George L. Germain. MA, CHCM, ASA, Executive Consultant for DNV-Loss Control Management, Loganville, GA.

quirements or standards.

3. Support, coaching and feedback.
4. Consequences.

Consequences are crucial. If a person's (or team's) performance is substandard, there should be some consequences associated with that performance. Likewise, if a person's (or team's) performance meets or beats the agreed upon standards, there should be some significant consequences associated with that performance. If there are no positive consequences, there is no motivation to continue that positive performance.

The importance of consequences is reflected in *Figure* 7-*11* "The ABC's of Human Behavior". The "A" (Activators or Antecedents) stimulates the "B" (Behavior), which results in "C" (Consequences, which tend to be either negative or positive). Behavior reinforcement strategy says that the consequences or results of what we do strongly influence whether or not we do it again. Negative consequences tend to decrease or stop the behavior that produced the consequences. Positive consequences tend to continue or increase the behavior that produced the consequences. In fact, the primary psychological basis for behavior reinforcement is that **Behavior is Influenced by Its Consequences.**

- Phase 6: On-Going Appraisal -

People are motivated to do what gets measured.

Continuous performance improvement requires continuous attention to performance monitoring - measuring - appraising. This applies to individual performance, team performance, or performance of the whole management system.

This chapter contributed by George L. Germain. MA, CHCM, ASA, Executive Consultant for DNV-Loss Control Management, Loganville, GA.

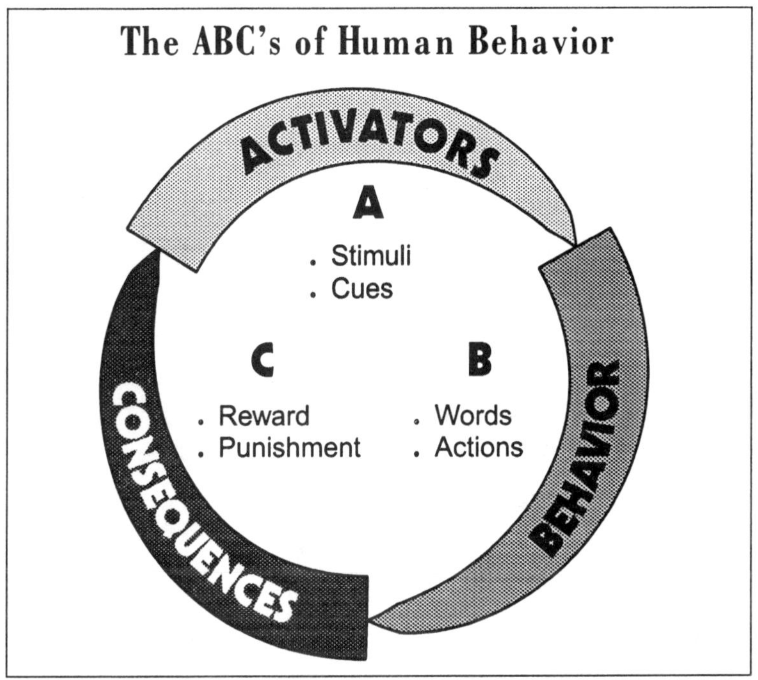

Figure 7-11

Regarding individual performance, here are seven appraisal tips:

1. EMPHASIZE PERFORMANCE, NOT PERSONALITY. Concentrate on the job objectives - job requirements - job performance - job results.

2. USE FACTS AND EXAMPLES. Try not to decide on the rating first and then dig up facts to support it. Try for facts and examples first, then the rating.

3. EVALUATE TYPICAL PERFORMANCE. Don't overemphasize one incident, either good or bad. Think of how the person usually performs on the factor being appraised.

This chapter contributed by George L. Germain. MA, CHCM, ASA, Executive Consultant for DNV-Loss Control Management, Loganville, GA.

4. AVOID ANCIENT HISTORY. Think in terms of how the person has performed for the past six months or year, and how she or he is doing now.
5. SHOW THE UPS AND DOWNS. Keep in mind that a person may do one part of the job well, another part just about average, and still another very poorly. Show the proper spread of performance levels.
6. DON'T "WHITEWASH." Show the facts as you see them and have them documented. It is important to know the individual's weak points or development needs, as well as his or her strong points.
7. KEEP COMPLETELY CONFIDENTIAL. Appraisal information is every bit as confidential as salary information or medical information. Treat it with equal respect.

Team performance can also be appraised in many different ways. One example is the "Team Effectiveness Assessment" shown in *Figure 7-12.*

Regarding the whole management system, a big question is "How well do we put performance measurement to work for safety?" How well do we go beyond measurements of consequences (injuries, damage, downtime and related losses), and get at the significant measurements of cause (both immediate and root causes)? How well do we go beyond measurements of cause and get at meaningful measurements of control (the leadership work that should be done)?

Measurements of *consequence* are after-the-fact, after the accident, after the loss. They reflect that we have a problem, but they tell us absolutely nothing about the nature of the problem, what is causing the problem, or what we can do about it. They are somewhat akin to a rear view mirror.

This chapter contributed by George L. Germain. MA, CHCM, ASA, Executive Consultant for DNV-Loss Control Management, Loganville, GA.

TEAM EFFECTIVENESS ASSESSMENT

Please circle one number for each item. The "5" means strongly agree, the "1" means strongly disagree and the other numbers range in between.

1.	Members understand their roles in the team.	1 2 3 4 5
2.	Members participate in setting team goals.	1 2 3 4 5
3.	Members are committed to working toward the goals	1 2 3 4 5
4.	Progress toward team goals is tracked.	1 2 3 4 5
5.	Members share responsibility for the work required to reach goals.	1 2 3 4 5
6.	Members are open in their communication with one another	1 2 3 4 5
7.	Discussions move along without significant concern for "sacred cows" or "taboo topics."	1 2 3 4 5
8.	Members feel free to express unusual or unpopular views.	1 2 3 4 5
9.	Conflicts are openly discussed and resolved.	1 2 3 4 5
10.	Members respect each other.	1 2 3 4 5
11.	Members trust each other.	1 2 3 4 5
12.	Members respect the leader.	1 2 3 4 5
13.	Members trust their leader.	1 2 3 4 5
14.	Members "pull for" one another.	1 2 3 4 5
15.	Strong and/or experienced members help those who are less capable and/or experienced.	1 2 3 4 5
16.	Members tolerate mistakes.	1 2 3 4 5
17.	Members can openly disagree with the leader.	1 2 3 4 5
18.	Members encourage experimentation and personal development.	1 2 3 4 5
19.	Members feel a sense of identity and pride in the team's work.	1 2 3 4 5
20.	The team identifies and solves work-related problems.	1 2 3 4 5
21.	Meetings are productive; objectives are accomplished.	1 2 3 4 5
22.	Members freely participate in discussions.	1 2 3 4 5
23.	Individual efforts are recognized within the group.	1 2 3 4 5
24.	Team efforts are recognized outside the group.	1 2 3 4 5
25.	The team effectively follows through the disposition of its recommendations and the results.	1 2 3 4 5

Figure 7-12

This chapter contributed by George L. Germain. MA, CHCM, ASA, Executive Consultant for DNV-Loss Control Management, Loganville, GA.

Guiding our safety system by measurements of consequence is like trying to drive down a highway by looking only in the rear view mirror... looking only backwards.

Measurements of *cause* add a significant dimension to program measurement. By getting at the "actual" and "potential" causes of accidents and incidents, we gain a great deal of insight into what must be done for prevention and control. Once we know the causes, especially the basic or root causes, we are well along the way to correcting them.

Measurements of *control* enable us to focus the spotlight where it can be most illuminating and do the most good... on answers to the question, "How well are we doing our management/leadership work for safety?" The International Safety Rating System TM: for example, enables measurement of the work being done in these twenty areas: Leadership and Administration - Leadership Training - Planned Inspections and Maintenance - Critical Task Analysis and Procedures - Accident/incident Investigation - Task Observations - Emergency Preparedness - Rules and Work Permits - Accident/incident Analysis - Knowledge and Skill Training Personal Protective Equipment - Health and Hygiene Control - System Evaluation - Engineering and Change Management - Personal Communications - Group Communications General Promotion - Hiring and Placement - Materials and Services Management Off-the-Job Safety.

> Continuous performance improvement
> requires
> continuous performance appraisal.

- Phase 7: Feedback & Coaching -

Following are seven key point tips for providing effective

This chapter contributed by George L. Germain. MA, CHCM, ASA, Executive Consultant for DNV-Loss Control Management, Loganville, GA.

performance feedback:

1. Provide on-going feedback on performance, problems and progress.
2. Use facts, figures, behaviors and incidents; get down to cases; discuss performance, not personality.
3. Pictures results with control charts, Pareto diagrams, check sheets, graphs and other pertinent visual aids.
4. Use questions; ensure two-way communication; learn to listen and listen to learn.
5. Discuss significant, timely pluses and minuses.
6. Seek basic causes and reasons for results.
7. Strive for agreement on how the person is performing, and why.

Here are ten important aspects of performance coaching:

1. Model desired performance.
2. Correct - commend - consult.
3. Mutually develop specific improvement plans.
4. Correct by reinstruction - reminders - reviews - refreshers - reinforcement.
5. Use punishment as a last resort.
6. Use a mutual problem-solving approach.
7. Base rewards on results and performance improvement.
8. Give immediate recognition for desired (efficient - safe - productive) behavior.
9. Emphasize attention, approval, assistance, success, satisfaction and support.

This chapter contributed by George L. Germain. MA, CHCM, ASA, Executive Consultant for DNV-Loss Control Management, Loganville, GA.

10. Make a habit of reinforcing positive performance, to make positive performance a habit.

Phase 8: Results and Recognition

Relating back to *Figure 7-10,* page 248, our performance management model for continuous improvement, the eight phases are:

Present Situation Appraisal
Dream, Vision, Mission
General Goals
S-M-A-R-T Objectives
Actions, Activities, Work
On-going Appraisal
Feedback & Coaching
Results and Recognition

When you do the first seven phases well, you can expect results such as these:

- Quality, production and safety statistics will improve, depending on the application. These can be measured easily by comparison with prior figures.
- As proper work procedures are followed to a greater degree, and by a larger number of employees, safety performance that controls injuries and property damage will show improvement... and can be measured by comparison with prior figures.
- Available production time will increase, since less time will be spent on "firefighting" a variety of problems caused by poor performance. Encouraging good performance takes less time than discouraging poor performance. Special efforts to "investigate poor quality," "discourage improper work procedures,"

This chapter contributed by George L. Germain. MA, CHCM, ASA, Executive Consultant for DNV-Loss Control Management, Loganville, GA.

and "stop growth of unsafe practices" will no longer require so much extra time. Concentration on promoting good quality, proper work procedures, and safe practices will achieve more effective results.

- Such things as maintenance costs, downtime, delays, and purchasing expenditures will improve as a reflection of improved work performance and better use of equipment, materials and people.
- Gripes, grievances and "groaning" will decrease as job satisfaction increases. Employees will develop a sense of pride in connection with their organization and their jobs.
- Team leaders and supervisors will be able to devote more of their time and talent to efficient and effective leadership work, since fewer "crisis" situations will erupt to consume their time.
- These kinds of results are stimulated and maintained by recognition — reward — reinforcement. Michael Le Boeuf, in his brilliant little book, *The Greatest Management Principle in the World*, phrased it very succinctly like this:

  ... establishing the proper link between performance and rewards is the single greatest key to improving organizations.

  ... Reward people for the right behavior and you get the right results.

A final point about the model. The process does not end with phase 8. As depicted in *Figure 7-10*, phase 8 leads right back to phase 1... and the cycle repeats and repeats for continuous performance improvement.

Key #5 - Persist and Repeat

So far in this section we have examined these four keys to

This chapter contributed by George L. Germain. MA, CHCM, ASA, Executive Consultant for DNV-Loss Control Management, Loganville, GA.

motivation management:

1. Clarify and communicate expectations.
2. Provide proper resources and support.
3. Emphasize education, training and development (in all three domains: attitudes, knowledge, and skills).
4. Practice performance management.

Though these strategies are *simple,* their implementation typically is *not easy.*

This brings us to key 5: PERSIST AND REPEAT. Implementation, maintenance and improvement of holistic motivation require work — repetition — work — persistence work!

Press on.
Nothing in the world can take the place of persistence.
Talent will not;
nothing is more common than unsuccessful men with talent.
Genius will not;
unrewarded genius is almost a proverb.
Education alone will not;
the world is full of educated derelicts.
Persistence and determination alone are omnipotent. .

The power of repetition is described in *Figure 7-13.* Read it now. Read every word of it. Read it and apply it to your on-going, never-ending motivational efforts. Apply it and reap the rewards of repetition and persistence.

Persist in applying the C-A-B (Cognitive — Affective — Behavioral) framework for holistic motivation.

Persist in applying these practical principles for managing motivation:

- The goals and objectives principle.
- The principle of involvement.

This chapter contributed by George L. Germain. MA, CHCM, ASA, Executive Consultant for DNV-Loss Control Management, Loganville, GA.

THE POWER OF SPACED REPETITION

By repeated listening, the subconscious mind becomes saturated with an idea. Soon the impact of the message brings a forceful change to the attitude of the listener. Thus, the ideas are received, retained, recalled and - most importantly - used!

THE POWER OF THE PILE DRIVER

If you've ever watched a pile driver at work, you've seen a weight of several tons come crashing down with terrific impact on the piling. Yet, the pile itself is driven down in fractions of an inch at a time. The process must be repeated hour upon hour, day upon day, before the piling is secure. The power of the pile driver, then, is not in its weight, not in the "punch" it carries, but in REPETITION!

So it is with the "power" of an idea! Seldom do we grasp an idea through sudden impact. Seldom are we overwhelmed by its weight. An idea or a concept grows with REPETITION. It becomes secure, firm, and deep-seated only after we have heard it again and again.

THE POWER OF "THE GREATEST WORDS..."

The most powerful ideas the world has ever known have come from the Bible. Yet there are fewer than 7,000 separate words used in both the Old and New Testaments! For two thousand years, and more, this Book has had more influence than all other ideologies combined. Several different writers, living in different centuries, in different areas, contributed to the Bible. But the ideas are the same. They are presented through REPETITION, REPETITION. One constantly-recurring theme runs through the Book, from Genesis to Revelation. It, too, has the power of the pile driver, because of REPETITION.

The power of an idea comes through repetition (it's received), repetition (it's accepted), repetition (it's retained), repetition (it's adopted), repetition (it's recalled) and repetition (it's applied).

THE POWER OF THE RAINDROP

In the arid southwest, a single raindrop can be a promise and a hope to the drought-stricken rancher. But the for "hope" to be realized, for the "promise" to be fulfilled, the single drop of rain must be RE-

This chapter contributed by George L. Germain. MA, CHCM, ASA, Executive Consultant for DNV-Loss Control Management, Loganville, GA.

PEATED over and over again, almost to infinity! (It takes some 75,000 raindrops to make one gallon of water . . . and it takes thousands of gallons of water to make a crop).

The life-giving qualities of the water are in each raindrop, but it takes REPETITION . . . REPETITION . . . REPETITION of the single drop of rain to sustain life!

North of Flagstaff, Arizona, the drops of rain fall on the parched earth, and by REPETITION, become first a trickle, then a stream, and finally a roaring torrent gouging its way into the face of the earth. REPEATED flow of the river becomes a superhuman sculptor, carving the statuary of the Grand Canyon, 1,000 . . . 3,000 . . . 5,000 feed deep into bedrock of Mother Earth. This is the power of the raindrop through REPETITION . . . REPETITION . . . REPETITION . . . REPETITION.

THE POWER OF A SINGLE SEED

In the Mariposa Grove of California's Yosemite National Park, there stands the oldest and most massive living thing on earth - the Giant Sequoia.

More than 300 feel tall, this majestic redwood tree measures 96½ feet in circumference, and weighs about 3,700 tons. Here, in simple truth, is living eternity; the infinite, burgeoning power of a single seed.

Scientists estimate that a tiny, wind-borne seed came to nest here about 3,800 years ago. Five hundred years passed. In faraway Africa, Akhenaton and Nefertiti ruled Egypt. The tiny seed had taken root and become a proud and sturdy redwood tree. It was already ancient when Moses led the Israelites into the Promised Land. It attained its full growth a thousand years later, while Socrates taught in Athens; and it reached its present towering majesty in the same century that Christ was born.

The seed of the Giant Sequoia (it takes 120,000 to weight one pound) is one of nature's "ideas."

Human ideas, yours or mine, possess the same potential strength and longevity. But to become as successful as the mighty redwood they must be nurtured, repeated, remembered and acted upon.

This chapter contributed by George L. Germain. MA, CHCM, ASA, Executive Consultant for DNV-Loss Control Management, Loganville, GA.

THE POWER OF COMMUNICATED IDEAS

The only discernible progress in the history of mankind has been the result of an IDEA combined with ACTION. An idea heard once and carelessly tossed aside benefits no one! But, with REPETITION . . . REPETITION . . . REPETITION . . . REPETITION, ideas become sharp, strong, powerful!

By repeated listening, the subconscious mind becomes saturated with ideas. Soon the impact of the message brings a forceful change to the attitude of the listener. Thus, the ideas are received, retained, recalled and - most importantly - acted upon!

Adapted from materials in the Professional Salesmanship Program, Success Motivation Institute, Inc., Waco, Texas

Figure 7-13

- The principle of mutual interest.
- The psychological appeal principle.
- The information principle.
- The principle of behavior reinforcement.

Persist in applying these four phases of the commitment process: Awareness—Activation—Affiliation—Assimilation.

Persist in applying these six ways to influence attitude changes:

1. Provide facts — knowledge — understanding.
2. Change their role or perspective.
3. Increase their involvement.
4. Attend to the affective aspects.
5. Model the behaviors that demonstrate desired attitudes.

This chapter contributed by George L. Germain. MA, CHCM, ASA, Executive Consultant for DNV-Loss Control Management, Loganville, GA.

6. Apply behavior reinforcement.

Persist in applying these five keys to motivation management:

Key #1 - Clarify and communicate expectations.

Key #2 - Provide proper resources and support.

Key #3 - Emphasize education, training and development.

Key #4 - Practice performance management...

... phase 1 - present situation appraisal.
... phase 2 - dream — vision — mission.
... phase 3 - general goals.
... phase 4 - S-M-A-R-T objectives.
... phase 5 - actions — activities — work.
... phase 6 - on-going appraisal.
... phase 7 - feedback and coaching.
... phase 8 - results and recognition.

Key #5 - Persist and repeat.

This chapter contributed by George L. Germain. MA, CHCM, ASA, Executive Consultant for DNV-Loss Control Management, Loganville, GA.

Selected References

Accident Prevention Manual For Business and Industry (Administration & Programs, 10th Edition). Itasca, IL: National Safety Council, 1992.

Bailey, Chuck. "Improve Safety Program Effectiveness With Perception Surveys." *Professional Safety.* October, 1993, pp. 28-32.

Bennis, Warren, and Burt Nanus. *Leaders: The Strategies for Taking Charge.* New York: Harper Collins, 1985.

Bird, Frank E., Jr. *Profits Are In Order*. Loganville, GA: International Loss Control Institute,1992.

Bird, Frank E., Jr. and George L. Germain. *Practical Loss Control Leadership*. Loganville, GA: Institute Publishing, 1990.

Bird, Frank E., Jr. and George L. Germain. *Commitment.* Loganville, GA: Institute Publishing, 1987.

Blitzer, Roy J., Colleen Petersen and Linda Rogers. "How To Build Self-Esteem." *Training & Development.* February, 1993, pp. 58-60.

Bradford, David and Allan Cohen. *Managing for Excellence.* New York: John Wiley & Sons, 1984.

Broad, Mary L. and John W. Newstrom. *Transfer of Training*. Reading, MA: Addison-Wesley Publishing Company, Inc., 1992.

Covey, Stephen R. *Principle-Centered Leadership*. New York: Simon & Schuster, 1990.

Daniels, Aubrey C. *Bringing Out The Best In People*. New York: McGraw Hill, Inc. 1994.

Drucker, Peter. *Management: Tasks. Responsibilities. Practices.* New York: Harper & Row, Publishers, 1973.

This chapter contributed by George L. Germain. MA, CHCM, ASA, Executive Consultant for DNV-Loss Control Management, Loganville, GA.

Dyer, William. *Contemporary Issues in Management and Organization Development*. Reading, MA: Addison-Wesley Publishing Company, 1983.

14 Elements of a Successful Safety & Health Program. Itasca, IL: National Safety Council, 1994.

Gardner, John. *Excellence.* New York: W. W. Norton & Company, 1984.

Hitt, William D. *Thoughts on Leadership*. Columbus, Ohio: Battelle Press, 1992.

Holcomb, Jane. *Make Training Worth Every Penny*. San Diego, CA: Pfeiffer & Company, 1994.

International Safety Rating System (Sixth Revised Edition). Loganville, GA: Det Norske Veritas, Inc., 1994.

Janicak, Christopher A. "Significant Accident Prediction." *Professional Safety*. July, 1994, pp. 20-25.

Johnson, W. G. *MORT: The Management Oversight & Risk Tree*. Washington, D.C.: U.S. Government Printing Office, 1973.

Kerr, Jeffrey and John W. Slocum, Jr. "Managing Corporate Culture Through Reward Systems." *Academy of Management Executive.* May,1987, Vol.1 No. 2, pp. 99-108.

Kinlaw, Dennis C. *Coaching For Commitment.* San Diego, CA: University Associates, Inc.,1989.

Kirkpatrick, Donald L. *Evaluating Training Programs: The Four Levels*. San Francisco: Berrett-Koehler Publishers, 1994.

Krause, Thomas R., Ph.D. and Kim C.M. Sloat, PhD. "Attitude Alone Is Not Enough." *Occupational Health & Safety.* January, 1993, pp. 26-31.

Paul, Karen B. and David W. Bracken. "Everything You Always Wanted To Know About Employee Surveys." *Training & Development.* January, 1995, pp. 45-49.

This chapter contributed by George L. Germain. MA, CHCM, ASA, Executive Consultant for DNV-Loss Control Management, Loganville, GA.

Petersen, Dan. *Techniques of Safety Management* (Third Edition). Goshen, NY: Aloray, Inc.,1989.

Petersen, Dan, "Establishing Good 'Safety Culture' Helps Mitigate Workplace Dangers." *Occupational Health & Safety.* July, 1993, pp. 20-24.

Planek, Thomas W. "Perception Equals Reality." *Public Risk.* January, 1994, pp. 14-16.

Pope, William C. *Managing for Performance Perfection.* Weaverville, North Carolina: Bonnie Brae Publications, 1990.

Quazi, Moumin. "Body And Spirit." *Occupational Health & Safety.* July, 1992.

Quick, Thomas L. *Unconventional Wisdom.* San Francisco: Jossey-Bass Publishers, 1 990.

Ramsey, Mildred. *The Super Supervisor.* Greenville, SC: Positive Presentations, 1986.

Sandler, Len. "Self-Fulfilling Prophecy: Better Management By Magic." *Training.* February, 1986, pp. 60-64.

Successful Manager's Handbook. Minneapolis, MN: Personnel Decisions, Inc., 1992.

Thomas, Michael C. and Tempe S. Thomas. *Getting Commitment at Work.* Chapel Hill, NC: Commitment Press, 1990.

Vogel, Christine. "Safety Surveys Build a Strong Foundation." *Safety & Health.* January, 1992.

Waitley, Denis and Reni Witt. "The Road to Success." *Possibilities.* September/ October, 1985, pp. 14- 17.

This chapter contributed by George L. Germain. MA, CHCM, ASA, Executive Consultant for DNV-Loss Control Management, Loganville, GA.

This chapter contributed by George L. Germain. MA, CHCM, ASA, Executive Consultant for DNV-Loss Control Management, Loganville, GA.

Chapter 8

The Cost of Accidents

"Not only is profit a measure of efficiency–good profits actually promote efficiency by enabling a business to plan and invest soundly for the future."

Frederick R. Kappel, Former President,
American Telephone & Telegraph

H. W. Heinrich, worldwide known author of the book, *Industrial Accident Prevention*, first published in 1931 by the McGraw-Hill Book Company, would unquestionably be at the top of most safety leaders' list of pioneer contributors to the cost and control of industrial accidents. While he probably did not coin the word or concept of direct - indirect accident costs, he is widely known to be one of the first to organize and quite clearly illustrate its use in safety literature as an effective motivational device with management. His occupational position as Superintendent of the Engineering and Loss Control Division of the Travelers Insurance Company placed him in a position to make studies of thousands of industry-wide accident statistics, their costs and control. This position also enabled him to be involved in many related subjects and develop a keen understanding of a broad range of industrial processes and practices.

Heinrich presents the subject of accident costs in 12 pages of his chapter, "The Basics and Philosophy of Accident Prevention." So inclusive is his listing of 11 items he called "Factors of Hidden Costs," commonly referred to as indirect or uninsured cost, that it compares quite favorably many cost references used in reputable studies as recently as 1995. Heinrich's original "Factors of Hidden Accident Costs" are shown in *Figure 8-1.*

It is interesting that Heinrich follows this early listing with the comment, "this list does not include all the points that might well receive consideration, although it clearly outlines the vicious and seemingly endless cycle of events that follow in the train of accidents." In another section of this chapter he goes on to state, "the incident cost of all non-compensable accidents is about one-half the total direct cost of all compensable accidents." These accidents are of a type not usually reported either to the state compensation board or to insurance compensation companies. Although *hidden accident-cost factors* have been specifically listed in this chapter, it should be emphasized that in calculating the 4:1 ratio between hidden and direct cost, only a few of the factors were employed. These being the ones for which data was readily available. His original research resulting in the 4:1 ratio was made through 1926. Though, there have been many variations of Heinrich's list of 11 factors of hidden costs.

Since Heinrich's publication in 1931, a literature search would indicate there have been well over 90 published articles or books discussing accident costs, with a significant

FACTORS OF HIDDEN ACCIDENT COSTS

1. Cost of lost time of injured employee.
2. Cost of time lost by other employees who stop work:
   a. out of curiosity.
   b. out of sympathy.
   c. to assist injured employee.
   d. for other reasons.
3. Cost of time lost by foreman, supervisors, or other executives as follows:
   a. assisting injured employee.
   b. investigating the cause of the accident.
   c. arranging for the injured employee's production to be continued by some other employee.
   d. selecting, training, or breaking in a new employee to replace the injured employee.
4. Cost of time spent on the case by first aid attendant and hospital department staff, when not paid for by the insurance carrier.
5. Cost due to damage to the machine, tools, or other property or to the spoilage or material.
6. Incidental cost due to interference with production, failure to fill orders on time, loss of bonuses, payment of forfeits, and other similar causes.
7. Cost to employer under employee welfare and benefit systems.
8. Cost to employer in continuing the wages of the injured employee in full after his return even though the services of the employee (who is not yet fully recovered) may for a time be worth only about half of their normal value.
9. Cost due to the loss of profit on the injured employee's productivity, and on idle machines.
10. Cost that occurs in consequence of the excitement or weakened morale due to the accident.
11. Overhead cost per injured employee - the expense of light, heat, rent, and other such items, which continues while the injured employee is a nonproducer.

Figure 8-1

number including emphasis on direct to indirect costs. While some do report this relationship as 1:1, many have reported ratios equal to or much greater than 4:1.

HISTORICAL RATIO REFERENCES.

1961 Bird and Germain. *Management Review,* 1:50 ratio.

1961 Wallach. indicated studies showing ratios from 1:1 to 20:1 Wisconsin Department of Labor-Industrial Relations, Landmark Study (1967), only one industry 4:1 in most cases 5:1.

1962 Crosby. Pit and Quarry Magazine cited studies ranging from 1:1 to 1:100.

1962 Hostletter.(Mining Congress Journal mining accidents far exceed 4:1 ratio.

1970 Fletcher and Douglas. *Total Environmental Control, 1:61 ratio.*

1993 HSE. British Study £1-£8 to £1-£50

See later comments for more current studies.

While there were many noteworthy contributions to the literature regarding accident cost through the years, much attention was focused on the costing method introduced into the published literature in 1963, by the Grimaldi and Simonds book, *Safety Management* (first edition). They presented a method of costing known as the "Modern Standard Method" that was widely accepted as more statistically accurate than previously used methodologies, such as simply multiplying the direct cost by 4.

Uninsured cost as referred to by Simonds and Grimaldi, was calculated by this basic formula. Uninsured cost equals A times the number of lost time cases, plus B times the number of doctor cases, plus C times the number of first aid cases, plus D times the number of no-injury accidents. A, B, C, and D are constants, indicating the average uninsured costs for each of the categories. Fatalities and permanent total disabilities were excluded, since these cases for most companies are so rare as to place them in a *catastrophe* category, subject to separate investigation.

In summary, this technique was probably the most practical method for the greatest number of safety leaders, especially those without the time, ready availability of data or resources to do their own research on the subject. A different method thought by many to be a more reliable method was thus presented for making calculations. Since this technique has been used by a number of undergraduate and graduate programs in safety for over two decades, adopted as the suggested method for establishing these costs in the 10th Edition manual of the National Safety Council's Accident Prevention Manual, it is no exaggeration to say it has *stood the test of time.*

Readers might conclude that there is only one method of calculating direct and indirect, or insured and uninsured, costs. However, we show that valid, verified studies have been made through the years and that accurate total cost of accidents can be determined in diverse ways. One recent study of accident cost was conducted in 1993 in the United Kingdom by the Government Agency (Accident Prevention Advisor Limited) that is the research wing of the Health and Safety Executive (HSE), a group with similar responsibilities to the National Institute of Safety and Health (NIOSH) in America.

We find it interesting from time to time to review the pioneering work of the past that paved the road to our present efforts to control these costly unintended events we call accidents. In this regard, we had to reflect on the early work of Mr. Heinrich. It continues to reflect great influence on our present day costing and control efforts in safety.

The remainder of this chapter is designed to add knowledge of both historical and current accident cost methodology, under the headings:

1. Property Damage, Huge Uninsured Cost.
2. 1993 HSE Accident Cost Study in United Kingdom.
3. The Cost of Risk.
4. Benefits of an Integrated System Approach to Safety/ EHS.

Although the one study was conducted in 1964, the authors feel the conclusions of this study has some important implications for the identification cost and control of the property damage accident in the future. We are sharing previously unpublished information on this costly accident virtually ignored in general industry throughout the world.

1. Property Damage, Huge Uninsured Cost.

You will remember that Mr. Heinrich identified his 5th factor of hidden accident cost as *cost due to damage to the machine, tools, or other property or to the spoilage of material.*

One of the authors started his career in Safety in 1951 at the Lukens Steel Company in Coatesville, Pennsylvania. Influenced by this #5 factor of Heinrich's hidden cost, he immediately began implementing an Accident Prevention Program aimed at the control of all accidents, causing harm to people or damage to property. By 1966, in a book titled, *Damage Control* co-authored by George L. Germain, an analysis of 90,000 accidents was reported, with 75,000 resulting in damage and 15,000 in injury. The ratio of uninsured to insured cost was 5:1. The ratio of accident types was 1 major or disabling injury to 100 minor and serious injuries to 500 property damage accidents. Lukens was following a proposed serious injury index suggested for use in the Z162 code at that time. Since the ratio of minor injuries to serious injuries was 12:1 at this time the Lukens ratio of 1-100-500 could also be represented as 1-7-38 since the 100 injuries included 12 serious injuries. This later ratio would seem to compare quite closely to Birds later 1969 Study while Director of Engineering Services at INA with a ratio of 1-10-30-600. His study of accident costs are also shown in the 1969 ratios, *Figure 8-2.*

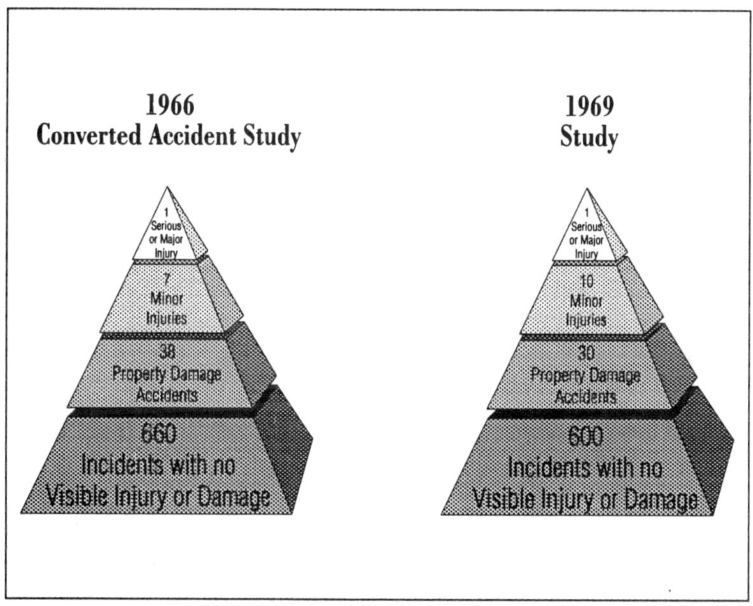

Figure 8-2

A visitation to the U.S. in 1962 was part of a tour of U.S. Steel plants, by two important representatives of the British Steel Industry, to determine if any new or different methods or techniques in accident control were being employed. While visiting Lukens as one of many steel plants, it was discovered that unlike all other plants visited, Lukens was employing a well developed total accident control program that included the property damage control as well as the personal injury approach common to other industrial plants at that time. This cost ratio was exhibited in both of Bird's 1966 and 1969 studies.

The British Steel Institute, upon the return of these two important representatives, promptly sent two researchers back to Lukens to audit the program, its cost and relationship to overall plant efficiency.

Since we feel that this important part of accident control is virtually ignored in the vast majority of U.S. companies, and since the cost of property damage is enormous, we feel it can be beneficial to review the original (1964) audit report by the British Researchers. Since property damage occurs at such a high rate cost, and is not properly dealt with by the majority of companies, we felt that anything to shed light on the reasons for its obvious neglect in most safety programs should be explored as we address the subject of *the cost of accidents*.

COST VERIFICATION OF 90,000 ACCIDENTS BY BRITISH STUDY TEAM AT LUKENS STEEL COMPANY IN 1963.

A visitation to the U.S. in 1962 was part of a tour of U.S. Steel plants, by two important representatives of the British Steel Industry, to determine if any new or different methods or techniques in accident control were being employed. While visiting Lukens as one of many steel plants, it was discovered that unlike all other plants visited, Lukens was employing a well developed total accident control program that included the property damage control as well as the personal injury approach, common to other industrial plants at that time.

In June 1964, Mr. Harry Van der Vord left Royal Society for the Prevention of Accidents (ROSPA) as head of its Industrial Safety Division to accept the post of Accident Prevention Advisor with the British Iron and Steel Federation

(BISF). Among other duties, he was particularly concerned with the study of accidental damage control. For this purpose he and a colleague, W. J. Shaw of the Human Factors Section of the British Iron and Steel Research Group Association (BISRA) came to Lukens Steel Company July 27, 1964, to thoroughly examine and analyze the pioneering damage control program that had been operating for over five years. Among the objectives of their study were the following:

a. to investigate the degree of savings achieved in damage cost since Lukens damage control system began.

b. to consider the relationship of damage control to other aspects of good management and its value as a measure of progress in this field.

c. to study in detail the structure and organization of the system.

d. to consider the techniques adopted to ensure the proper functioning of the scheme, and its effect on the efficiency and administrative work.

e. to sample the opinions and attitudes of those involved with the scheme, administrative burden and to consider the means adopted to ensure support.

Conclusions quoted directly from the final study report of Msrs., Van der Vord and Shaw submitted to the BISF:

A. "The claims of savings achieved in damage cost at Lukens are valid. The reporting of damage incurred if not complete, approaches the target of completeness and the costing contained in damage control records, though at present done independently and in parallel to the companies accounting system are sufficiently accurate and reliable. A check on the costing in a sample of actual damage cases revealed only insignificant errors of under-calculations. A further broad check, with the records and costing maintained by the company's maintenance operation control, confirmed that the damage control estimate of the total damage cost are, if anything conservative."

B. "There is no statistically significant correlation between the reduction in damage cost since 1959, the base year of the Damage Control Scheme, and variations in the Disabling Accident Frequency and the All Accident Frequency Rates. A Serious Injury Frequency Rate, introduced in 1960, and considered to be a truer indication of hazards within the Works, shows a 50% reduction by 1963 in concert with the reduction in damage cost recorded. There has, however, been no step by step reduction, as occurred with damage cost, and the statistical correlation is significant only at the 10% level.

Examination of the figures within departments and, for seasonal variations, between the 13 periods of the fiscal year again reveals no correlation between injury and damage experience.

It is possible that the causes of the more costly damage highlighted by the Scheme are not closely associated with potential injury. If so, a reduction of injury rates in parallel to the early dramatic savings in damage cost is not to be expected. When the Scheme was begun at Lukens the Disabling or Lost Time Injury rate was already at a level which, by any standard of comparison, must be considered fantastically low. A significant impact on this rate would be difficult to demonstrate statistically. We consider nevertheless that the Damage Control Scheme is playing an important part in the concerted efforts of management to create a climate of efficiency within which a further lowering of injury rates may yet be achieved. Though the direct connection would be impossible to establish, we are also persuaded that the Scheme is at least contributing to maintain the outstandingly low rates already achieved."

C. "We heard repeated evidence of management's belief that Safety, Quality, Productivity and Cost Reduction are not contradictory or separate ends to be pursued in isolation, but merely different facets of the single drive towards efficiency. Management certainly regards the Damage Control Scheme as an inseparable part of this wider approach. The investment of money and effort in the Scheme can be amply justified in terms of damage cost reduction alone but its influence probably extends, through this concerted drive for efficiency, to fields beyond those of accidental injury and damage solely. Other indices of efficient performance are maintained by the Company including a percentage figure for quality rejects, a percentage figure for orders shipped on time, and a "work performance" index. We have studied variations in these indices in relation to the Damage Severity rates and find that a close correlation exists. This bears out the view of Lukens management that the Damage Control Scheme far from hindering production, is an inseparable part of the total effort for efficiency.

D. "We consider that the Damage Control Scheme at Lukens effectively reduces damage and contributes to safety, and that by following in principle the method adopted there, such a scheme could be started anywhere, probably with existing staff. Full details of the structure and organization of the scheme are therefore given in the body of this report.

E. "Efficient reporting is the mainstay of damage control. In theory, the obvious drawback to efficient reporting is the reluctance on the part of the operator to involve himself or others in blame. Lukens have overcome this (i) by firm instructions backed by top management that reports *must* be made (ii) by daily effective control at the repair point and on the works, to detect unreported damage. There is little or no chance of avoiding the system and to fail in reporting may involve further blame. As far as 'tale-telling' is concerned, the blame is transferred to the system by the techniques employed. The duties of the Damage Control Inspector play a considerable part in the efficient functioning of the scheme by preserving a practical outlook throughout.

The administrative burden is not heavy. At present, at Lukens, the volume of paper work has been reduced by (i) a precise and selective definition of material damage which reduces the number of reports to be handled. (ii) by on-the-spot estimating of the cost of the damage in each report from easily accessible cost data. (The verification of damage estimates may shortly be improved by a simple feedback of actual cost from the Maintenance Control Department of the Company)."

F. "Every opportunity was given for uninvigilated discussion with workers and supervision. Each one interviewed knew the Company's policy towards accident prevention and damage control and understood his own responsibilities. The good communications observed must contribute towards this close understanding. Safety training has played an obvious part in improving communications.

With the problem of reporting largely overcome, the work force is participating in the revelation and assessment of damage. They have become both cost and damage-conscious through this participation. It is logical to suppose that this awakened interest, by itself, contributes towards accident reduction.

There was no evidence of tiresome administrative burden on the supervision. There were grounds for believing that the concept of damage control has been accepted as a sensible workaday arrangement and worthy of co-operation. The techniques adopted to ensure support of the scheme can be said, with justification, to have succeeded."

We have included this historical perspective of the Lukens Property Damage study to focus much greater attention on this hidden cost, which we believe to be one of the largest, yet almost totally ignored by most organizations. The reporting of an injury establishes a work-related occurrence and is primary evidence for the employers responsibility for coverage by workers compensation. This is probably the biggest reason employees can be encouraged to report injuries quite readily.

On the other hand, the reporting of property damage, tends to invite blame and discipline. Thus people are not as apt to report damage as they are to report injuries. It's quite interesting that key maintenance executives use the cost of replacing or repairing the previous year's accidental property damage as an important component of their departmental budget for the following year. One of the frequent criteria to appraise a leaders performance is to evaluate how well he/she controls budget. The upper limit of control is frequently the budgeted cost. It's quite interesting that 10-30% of an organization's maintenance costs are controllable accident damage and yet by the very managing of property damage, within the limits of control, the leader may add to his recognition as a good performer.

2. 1993 HSE Accident Cost Study in United Kingdom.

A more recent extensive 1993 study, "The Cost of Accidents at Work" was reported by the Health and Safety Executive Group of the British government in a 1993 publication by the same name. The study was conducted by a team of professionals that included an economist. Interestingly enough, part of the published report was a ratio of major and over-3 day injury accidents to minor injuries to non-injury accidents (*Figure 8-3*). Studies were conducted

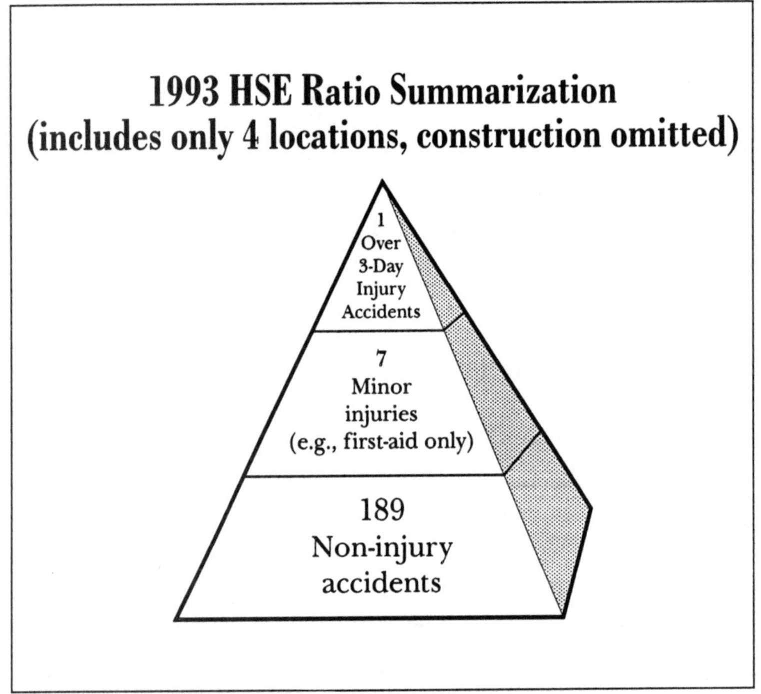

Figure 8-3

at five different types of industrial locations that included: 1. a construction site, 2. a creamery, 3. a transport organization, 4. an oil platform and 5. a hospital.

Some interesting facts about the study at each location follow:

1. Construction
   Study period in weeks - 18
   120 working on-site
   Over 3-day lost time injuries - 0
   Minor injuries - 56
   Non-injury accidents - 3570
   Total Accidents - 3626
   Injury : non-injury accident ratio - 1:64
   Accidents/year/employees - 87
   Insured to uninsured cost ratio - 1:11

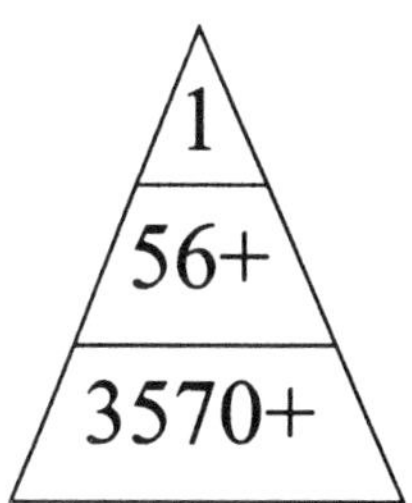

2. Creamery
   Study period in weeks - 13
   338 working on-site
   Over 3-day lost time injuries - 6
   Minor injuries - 31
   Non-injury accidents - 889
   Total accidents - 926
   Injury : non-injury accident ratio - 1:24
   Accidents/year/employees - 11
   Insured to uninsured cost ratio - 1:36

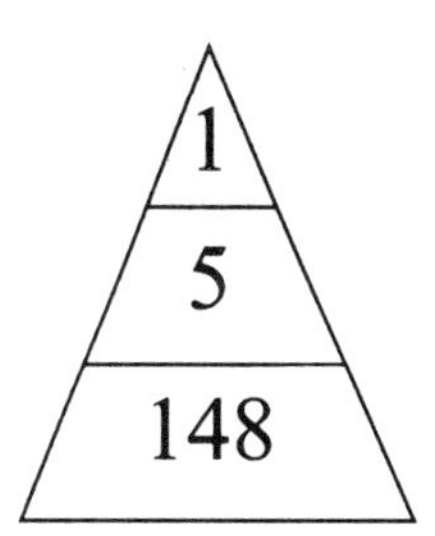

3. Transport
   Study period in weeks - 13
   80 working on-site
   Over 3-day lost time injuries - 0
   Minor injuries - 0
   Non-injury accidents - 276
   Total accidents - 296
   Injury : non-injury accident ratio - NA
   Accidents/year/employees - 74
   Insured to uninsured cost ratio - 1:8

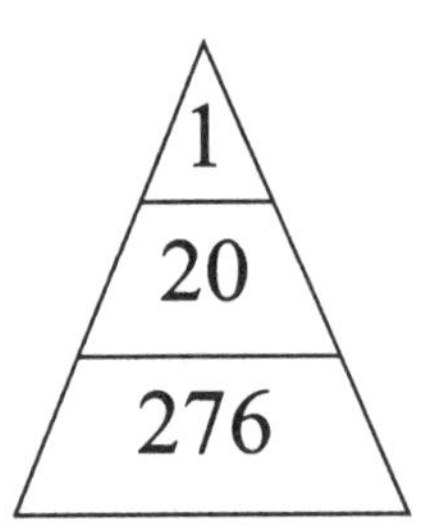

4. Oil Platform
   Study period in weeks - 13
   210 working on-site
   Over 3-day lost time injuries - 2
   Minor injuries - 8
   Non-injury accidents - 252
   Total accidents - 262
   Injury : non-injury accident ratio - 1:25
   Accidents/year/employees - 5
   Insured to uninsured cost ratio - 1:11

1
4
126

5. Hospital
   Study period in weeks -13
   700 working on-site
   Over 3-day lost time injuries - 6
   Minor injuries - 58
   Non-injury accidents - 1168
   Total accidents - 1232
   Injury : non-injury accident ratio - 1:18
   Accidents/year/employees - 7
   Insured to uninsured cost ratio - na

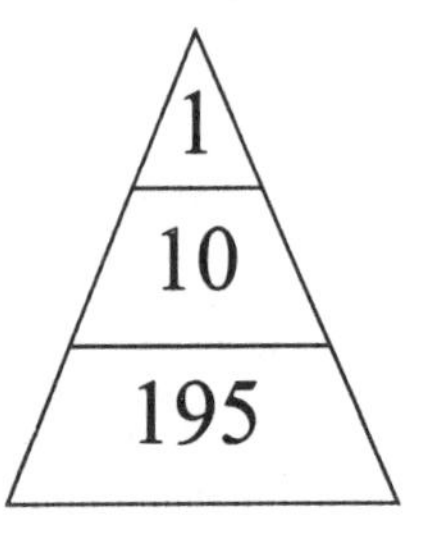

6. All Studies
   Study period in weeks - 70
   1448 working on-site
   Over 3-day lost time injuries -14
   Minor injuries - 153
   Non-injury accidents - 6175
   Total accidents - 6342
   Injury : non-injury accident ratio - 1:37
   Accidents/year/employees - 17
   Insured to uninsured cost ratio - na

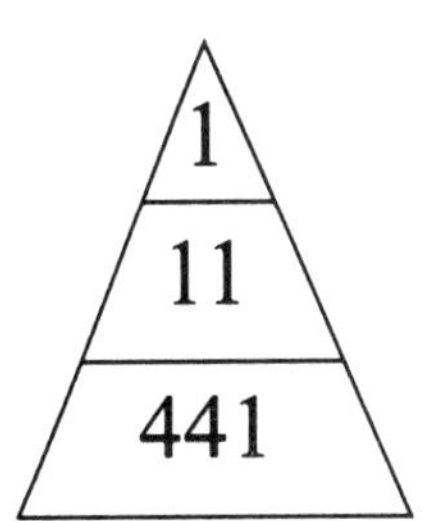

7. All Except Construction
   Study period in weeks - 52
   1328 working on-site
   Over 3-day lost time injuries -14
   Minor injuries - 97
   Non-injury accidents - 2605
   Total accidents - 2716
   Injury : non-injury accident ratio - 1:23
   Accidents/year/employees - 8
   Insured to uninsured cost ratio - na

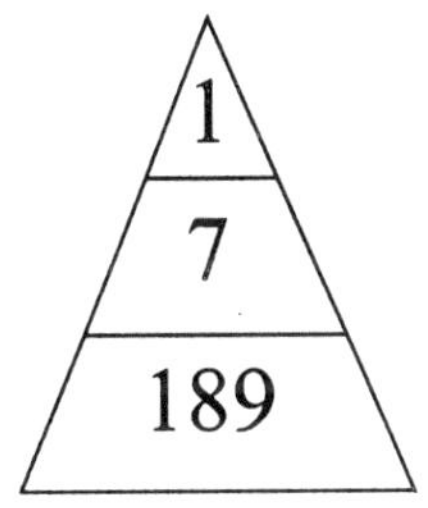

The iceberg summarization of insured, uninsured cost of 4 locations, not including construction, is shown in *Figure 8-4.*

Some of the interesting facts and results of HSE accident cost study should create considerable attention of industrial leaders around the world.

Very Interesting Findings

- results reflect actual costs recorded as they occurred; estimates were not used.

- uninsured costs far exceeded insured costs.
- one organization's costs were as much as 37% of its annual profits.
- another's accident costs were equivalent to 8.5% of its tender price.
- a third organization's costs were equivalent to 5% of its running cost.
- losses result primarily from failures of management control, and when not controlled can interact and escalate.
- a comprehensive program of accident prevention is necessary to achieve success in occupational health and safety management. Focusing on personal injury accidents is not enough. There needs to be practical management control programs which control all potential loss.
- fundamental to efficient control of loss is the elimination of causes to non-injury accidents at the base of the triangle.
- while there is a wide range of immediate causes of accidents, the underlying causes are common.
- a separate analysis of 80% of the accidents showed that over 8% were judged to have the potential for serious consequences such as fatalities, multiple injuries or catastrophic loss.
- it was found that focusing on the prevention of reported personal injury accidents is not enough. There needs to be proactive management control programs which prevent or control all potential sources of loss.

Isn't it completely understandable why executive leaders in the United Kingdom would make the following state-

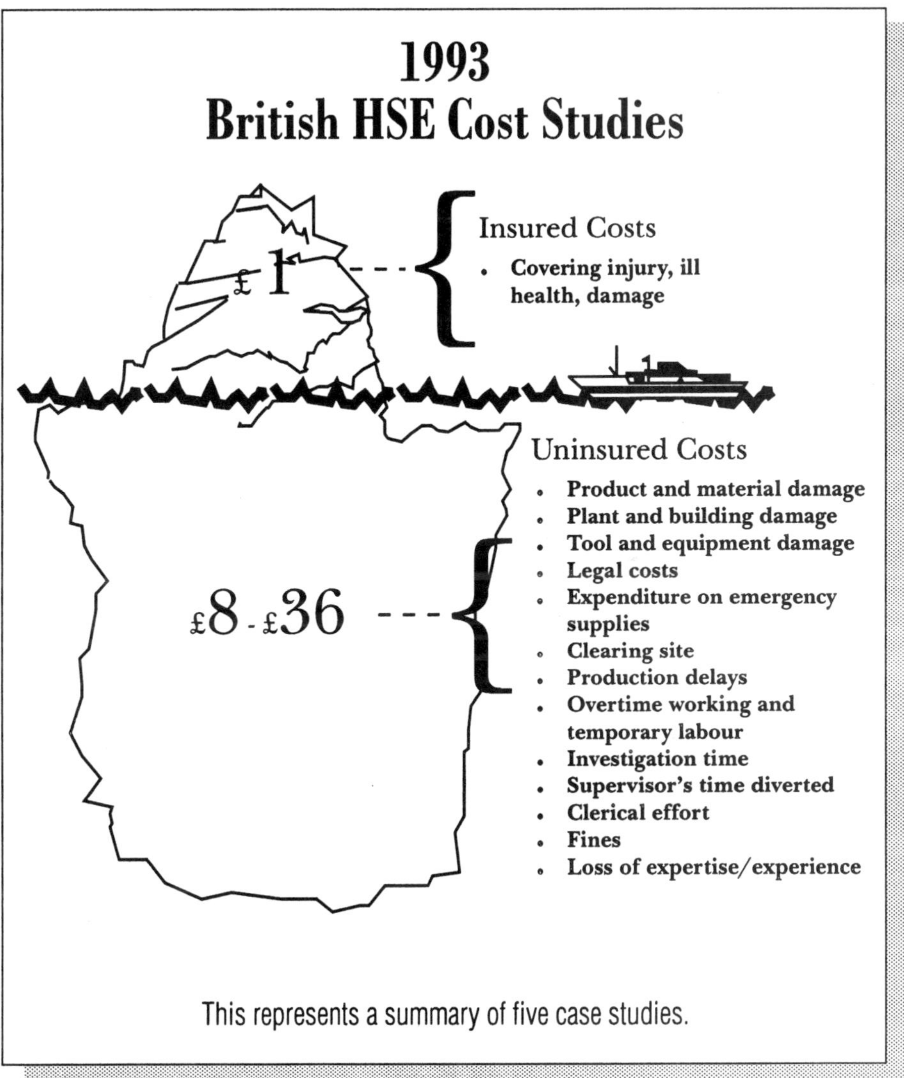

Figure 8-4

ments? (Quoted from the HSE's "The Costs of Accidents at Work.")

> "We recognize the importance of costing loss events as part of total safety management. Good

safety is good business."
—*Dr. J. Whiston, ICI Group Safety, Health and Safety Manager*

"Safety is, without doubt, the most crucial investment we can make. And the question is not what it costs us, but what it saves."
—*Robert E. McKee, Chairman and Managing Director, Conoco (UK) Ltd.*

"At Shell, we believe that an excellent company is by definition a safe company. Since we are committed to excellence, it follows that minimizing risk to people, plant and products is inseparable from all other company objectives."
—*Shell Exploration and Production*

"Prevention is not only better, but cheaper than cure ... There is no necessary conflict between humanitarian and commercial considerations. Profits and safety are not in competition. On the contrary safety at work is good business."
—*Bask Butler, Managing Director, British Petroleum Company, plc*

3. The Cost of Risk

When addressing the subject of accident cost, the typical safety leader will invariably think of the cost of his/her organization's workers' compensation. Seldom is there ever any reference to the huge accident cost categorized and financially handled as property or liability loss coverages. Increasingly safety leaders of major organizations will more accurately address their accident cost from their Cost of Risk summary. At a recent group meeting, we heard a corporate Director of Environment, Health and Safety refer to Cost of Risk for the first time in several years.

The question is why? The answer is that the vast majority of safety personnel are not acquainted with the term Cost of Risk. Risk as used in this book is "chance of loss." Perhaps an even sadder fact is that Cost of Risk concept is seldom taught, or discussed in most Safety Management Training Programs. The term "Risk" in the context of this discussion means "Chance of Loss," and particularly from the pure risks of business. In the management of risk there are two types most frequently referred to. The pure risk is one that can only result in loss or no loss but never gain. The speculative risk is one that can result in gain or loss. This is the typical business risk taken regularly to improve efficiency.

It is our opinion that in order to establish a summary of all costs of accidents for your organization, or the total costs resulting from the pure risks of business, you absolutely must know your Cost of Risk.

It was June 14, 1962, that a pioneer in risk management, Douglas Barlow, a member of the bar in the Province of Quebec, Canada and manager of Worldwide Risk Management services for the Massey-Ferguson Corporation, made a landmark address concerning the costing of risk, before the Insurance Buyers Association of Toronto. Mr. Barlow opened this now historic presentation with insightful personal thoughts on where the risk manager fits in the organization and what he/she was trying to do. What is the job? After emphasizing the practical importance of knowing this both for your company and yourself, he went on to say that our companies are exposed to risks of loss, most of which are called "accidents" and our fundamental task is to control them. Too much of the time we lose sight of the forest because of the trees. In effect, we fail to properly evaluate all the related costs of destruction of resources, the resultant liabilities and related costs.

By failing to know more about Cost of Risk we fail to identify the total cost of our accidents. Thus we lack the powerful motivational force, that would help gain the commitment of key executives and encourage control of the risks causing them. In effect, we fail to make the maximum contribution to our company's prime reason for existence.

> "Profit is the ignition system of our economic engine."
> —Charles Sawyer

Shortly before presenting his now widely used Cost of Risk concept, Barlow used the following words which are as appropriate now as they were then. "The test of a company's operations is the dollar figures in its annual statement of profit and loss and assets and liabilities. Any cost reduces both income and net worth, risks of accidental loss are one of the costs of operations. Uncontrolled, this cost is unpredictable in impact, a menace to earnings and worth, even insolvency. It may be brought under control - minimized over the years, measurable enough that account may be taken of it in planning and budgeting. This is risk management, expressed in dollar terms. These are the terms that are used to express and test the end result of the entire operations of the company; they seem, therefore the most appropriate for expressing and testing risk management."

Mr. Barlow then presented the Cost of Risk concept that follows:

"THE COST OF RISK"

1. "cost" direct and consequential measures for loss prevention, plus
2. insurance premiums,
3. losses sustained (including consequential effects) and

PURE RISKS

CHANCE OF LOSS FROM:

flood, fire, injury, fraud, stress, pollution, sabotage, tornado, snow damage, explosion, earthquake, riot, drug abuse, theft, hijacking, homicide, property damage, cargo damage, storm damage, terrorism, fire, explosion, injury, illness, stress, alcoholism, equipment or cargo sabotage, hail damage, utility breakdown, fleet damage, product liability, wind, damage, environmental harm, environmental damage.

immediate expenses to curtail losses, minus

4. payments by insurers, to or on behalf of the assured, minus
5. recoveries from third parties, plus
6. administrative expenses.

All of these factors are in some degree controllable, and some have an affect upon others. To balance them so as to get as close as possible to the desired result is our job."

The Cost of Risk refers to all costs associated with an organization is risk management function including risk control and risk financing. There has been very little change in what constitutes Cost of Risk since Mr. Barlows day, even though descriptions may vary slightly from one individual to another. One of Americas most respected authorities on Risk Management, George L. Head, PhD., CPCU, ARM, CSP, CLU, Vice President of the Insurance Institute of America has co-authored the widely used book, *Essentials of Risk Management*, with Stephen Horn II, CPCU, ARM, AAI. In the books Second Edition 1991, they present a definition of Cost of Risk as a total of the following:

1. The cost of accidental losses not reimbursed by insurance or other outside sources
2. Insurance premiums or payments to other outside sources of funds
3. The costs of measures to prevent or reduce the size of accidental losses
4. The administrative costs for risk management.

Figure 8-5 shows how "Cost of Risk" fits into the Risk Management process.

Authoritative Accident Costing Sources

Perhaps the three most respected authorities on accident costs here in the U.S., if not in the world, would be the National Safety Council (NSC), the Tillinghast-Towers Perrin organization and the Risk and Insurance Management Society. The NSC has been the publishers of the well-known *Accident Facts* publication for many years with the cooperation of many city, state and federal organizations and agencies involved with related statistical data. Through a cooperative effort by Tillinghast-Towers Perrin and the Risk and Insurance Management Society, a Cost of Risk Survey has been made each year since 1977 and published in a book by the same name.

While comments will be made concerning data presented in both publications, the quotations and comments that follow immediately have been take from the 1995 edition of the *Accident Facts* publication by the National Safety Council. The "Accident Facts" publication points out that "the true cost to the nation, to employers and to individuals of work-related deaths and injuries is much greater than the cost of workers' compensation insurance alone." The figures presented below show the NSC's estimate of the total costs of occupational deaths and injuries.

TOTAL COST IN 1994 OF WORK-RELATED DEATHS AND INJURIES —$120.7 BILLION

Please note: The very important explanation of costs that appear directly below this figure includes wage and productivity losses of $60.9 billion; medical costs of $21.1 billion, and administrative expenses of $23.7 billion. Including employer costs of $10.6 billion such as the money value of time lost by workers other than those with disabling injuries, who are directly or indirectly involved in injuries, and

RISK MANAGEMENT FLOWCHART
RISK IDENTIFICATION
EXPOSURE ANALYSIS
RISK CONTROL PLAN
AVOIDANCE
LOSS PREVENTION
LOSS REDUCTION
RISK CONTROL EXPENSES
RISK FINANCING PLAN
RETENTION
TRANSFER
LOSSES
EXPENSES
PREMIUMS
BORROWING
COST OF RISK
MONITOR RESULTS
ACCEPTABLE
UNACCEPTABLE

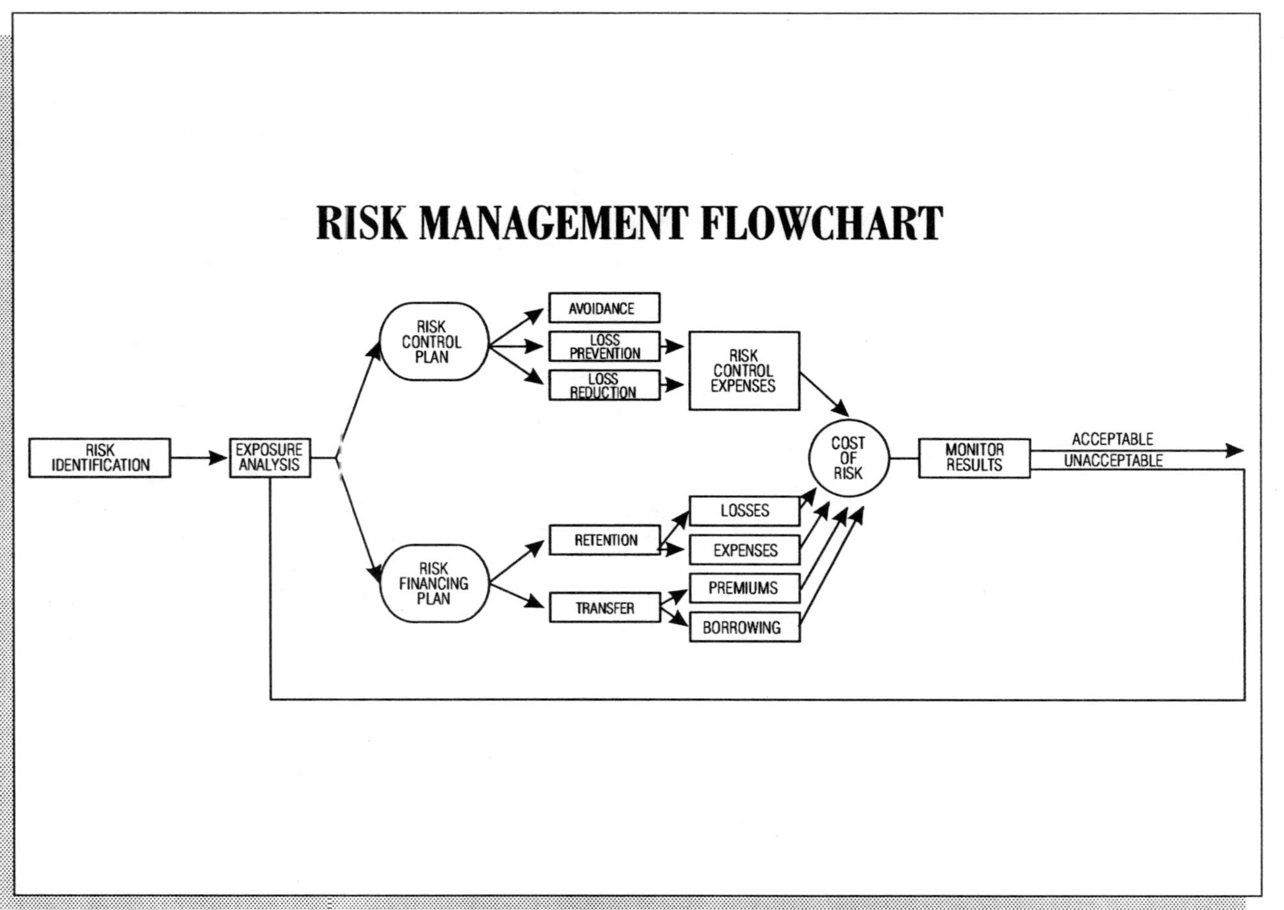

Figure 8-5

the cost of time required to investigate injuries, write-up injury reports, and so forth. Also includes damage to motor vehicles on work injuries of $2.0 billion and fire losses of $2.4 billion.

Note that a large portion of the $120.7 billion are costs other than workers' compensation. As a matter of fact, its rather amazing that in discussions with numerous safety personnel, the preceding paragraph, very clearly shows that a thorough description of total costs, has often been overlooked. This is one of those situations where readers look but do not see. Many, therefore, miss the very important explanation that a number of costs referred to earlier in this chapter as indirect or uninsured costs are included to make up a substantial body of the $120.7 billion total costs. Careful reading of the paragraph explaining total costs will also point out in the first sentence that the $120.7 billion includes costs to the nation, to employers, and to individuals.

Because the use of accident costs as a motivational device to gain increased commitment by leaders is considered so important, a more detailed explanation of the NSC's rationale in using uninsured costs printed in the Appendix of this book.

In presenting the total cost of work-related deaths and injuries as $120.7 billion, the NSC clearly states that it is their estimate. We suggest that this estimate is conservative. It does not include the high product, professional or general liability costs that are related to the accidents. In addition, property risk financing costs such as those related to business interruption, boiler and machine, fire, and so forth, significantly influence the broader Cost of Accidents as compared to the exclusivity of death and injury. This is one of the big reasons for EHS personnel to show more interest in the "Cost of Risk."

The Executive Summary that follows is quoted directly from the 1955 publication, *Cost of Risk Survey* published by Tillinghast-Towers Perrin Risk Management Publications.

WHAT IS THE "COST OF RISK?"

"In 1962 the concept of the cost of risk was developed by Douglas Barlow, a former risk manager and president of RIMS, to include the following elements: net insurance premiums, retained losses, risk control and loss prevention expenses, and administrative costs. "Cost of risk" thus refers to all costs associated with an organization's risk management function, including risk assessment, risk control, risk financing, and risk administration.

The cost of risk concept was further developed in 1993, when the Institute of Management Accountants and RIMS published a Standard of Management Accounting Statement entitled "Practices and Techniques: Internal Accounting and Classification of Risk Management Costs." The goal of the Statement was to provide risk managers with a consistent and comparable method of accounting for risks. It defines risk management costs as applying to three exposures: property, tort liability, and occupational disease or injury. Within the three exposures are six types of costs: insurance premiums; retained losses; internal administration; outside services; financial guarantees; and fees, taxes, and other similar expenses.

The Cost of Risk Survey follows the standards in

the Statement with the exception of risk control expenses. Risk control includes such functions as employee safety, fire protection, environmental affairs, fleet safety, security, emergency and contingency planning, and product safety. Some capital expenditures, such as sprinkler systems, might also be considered "risk control: by some companies. risk control activities are the responsibility of many departments within an organization, which complicates data gathering. Furthermore, it has proven difficult for survey respondents to differentiate risk control expenses from the normal cost of doing business. For this reason, this element of the traditional definition of the cost of risk has been excluded form the cost of risk survey questionnaire since 1990.

For purposes of this report, the definition of cost of risk includes the following components:

- insurance premiums for liability, property, and workers' compensation exposures, including any associated taxes
- retained/uninsured losses for liability, property, and workers' compensation exposures, including any associated financial guarantees, fees and taxes
- risk management/insurance department administrative budget
- costs for outside services, including broker/ agent fees, outside consulting and actuarial fees, risk management information systems, and captive management costs."

Many EHS professionals do not realize that workers' compensation is only 40.2% of the cost of risk with property coverage at 18.4% and liability at 36%. You will note by looking at *Figure 8-6* that the trend to retaining risk is clearly well on its way at 52%.

The positive trend to retained risk sends a loud and clear message to all EHS personnel. We must all acquire a much better knowledge of the Cost of Risk in order to (1) recognize we are an important part of the risk management team and the importance of our assessment of all risk to decisions that directly effect the bottom line (2) we must maintain an improved interaction with risk finance personnel in order to optimize the effectiveness of risk management decisions, (3) know the true value of cost of risk trends to the evaluation of the loss control effectiveness.

Let's use a few of the statistics from the "Survey" to get the big picture of costs. Using the "Corporate Scoreboard" of sales and profits from the March 5, 1996 issue of *Business Week*, in consort with our cost of risk data from our Tillinghast Survey we determine the following:

We caution readers that, determining the estimated cost of risk for your company, by utilizing the percentage of revenue may not provide the highest degree of executive motivation.

By dividing the cost of risk as a percentage of revenue into your organization's net profit dollars for the year, you will have the percentage of profit used to pay the cost of risk for your organization. This figure will grab the attention of every leader from the floor to the boardroom. Let us look at this application utilizing entire industrial sectors.

COST OF RISK SURVEY

12 Months Sales 1994		Average Cost of Risk as a Percentage of Revenues 1994 .73%
$ 4,416,362.7	$ML	$ 32,239.4

COST OF RISK AS A PERCENT OF PROFIT

12 Months Profit 1994 $ML	Cost of Risk as a Percentage of Profit 12.7%
$ 252,279.4	$ 32,796.4

COST OF RISK FOR SPECIFIC INDUSTRIAL SECTORS

Industry	Revenues	Percent of Profit
Healthcare	$4,315,000,000	9%
Transportation	$141,300,000,000	3.1%
Electric & Electronics	$154,650,000,000	7.3%
Chemical	$136,389,000,000	6.2%

Two of the charts on the next pages that follow show the total cost of risk as a % of Revenue for the manufacturing and service industries in the US a third chart shows the composition of risk dollar by Revenue size. The 121 page "Cost of Risk" survey publication contains a number of graphs or charts. All data contained in the survey was obtained from respondents cooperating. (See Appendix 1).

By now it should be quite evident that lowering this cost of accidents and risk in general is one of the few remaining areas that remains in the management system for substantial cost reduction.

There are however, several very important related points that need to be recalled to the readers attention.

1. The Cost of Risk Survey publication points out that it follows the standards repeated below for your convenience from page 24.
    - insurance premiums for liability, property, and workers' compensation exposures, including any associated taxes
    - retained/uninsured losses for liability, property, and workers' compensation exposures, including any associated financial guarantees, fees and taxes
    - risk management/insurance department administrative budget
    - costs for outside services, including broker/agent fees, outside consulting and actuarial fees, risk management information systems, and captive management costs
2. The reason for the exclusion of the $ of risk control is also presented on page 26 and again for the readers convenience printed here:

    "Risk control activities are the responsibility of many departments within an organization which complicates data gathering. Furthermore, it has proven difficult for survey respondents to differentiate risk control expenses from the normal cost of doing business. For this reason, the element of the traditional definition

Composition of the Cost of Risk
1.8
Outside
Services
2.9
Risk Management
Department Costs
10.6
Property
Premiums
7.8
Retained Property
Losses
27.8
Retained
Workers'
Compensation
Losses
20.1
Liability
Premiums
12.4
Workers'
Compensation
Premiums
16.6
Retained
Liability
Losses
1994
The Cost of Risk Dollar
13%
Property Loss
36.7%
Liability
Losses
40.2%
Workers'
Compensation
Loss
2.9%
Risk
Management
Department
Costs
1.8%
Outside
Services

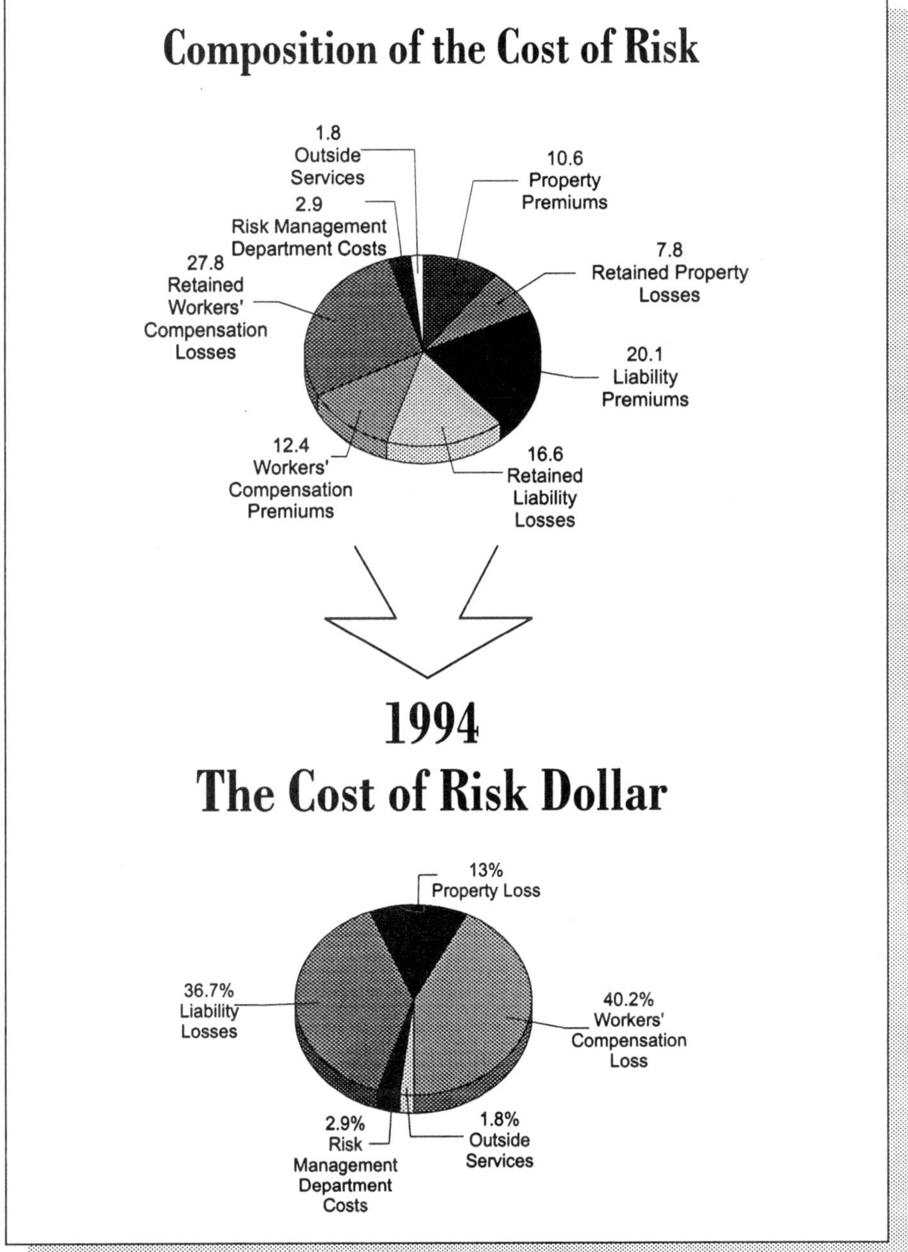

Figure 8-6

Total Cost of Risk as a % of Revenue
U.S.-Based Manufacturing Industries
Mining & Energy
1.11%
1.39%
Primary Metals, Leather & Stone
1.00%
0.88%
Metal Products
1.02%
0.85%
Lumber, Furniture & Packaging
0.51%
0.79%
Transportation Equipment
0.99%
0.79%
Machinery
0.70%
0.73%
Construction
0.97%
0.72%
Food, Agriculture, Tobacco, Textiles
0.40%
0.66%
Misc. Manufacturing Industries
0.74%
0.64%
Chemicals, Rubber & Plastic
0.51%
0.52%
Electronic and Other Electrical Equipment
0.38%
0.32%
Printing and Publishing
0.41%
0.32%
Average, All Manufacturing Industires
0.60%
0.62%
1993
1994
1995 Cost of Risk Survey - Tillinghast-Towers Perrin Organization

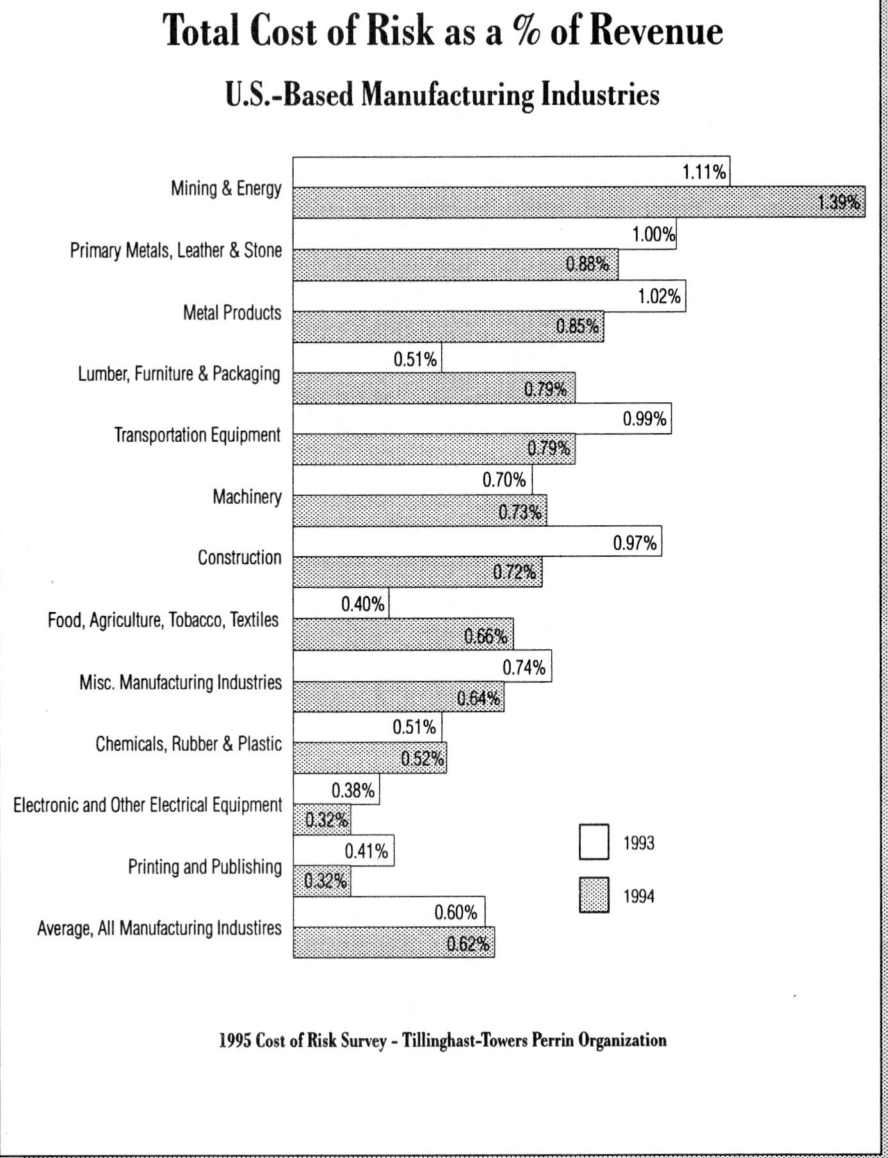

Figure 8-7

Total Cost of Risk as a % of Revenue
U.S.-Based Service Industries
Transportation Service
3.37%
2.70%
Health Care
2.02%
2.03%
Real Estate, Security/Commodity Brokers
1.29%
1.92%
Governmental
1.20%
1.20%
Personal, Business Service
1.84%
1.20%
Educational, Nonprofit Institutions
0.83%
1.10%
Retail Trade
0.85
0.67%
Combination Utility
0.75%
0.66%
Telecommunications
0.54%
0.65%
Electric Utility
0.64%
0.63%
Natural Gas Utility
0.82%
0.38%
Wholesale Trade
0.48%
0.32%
Insurance
0.17%
0.20%
Finance-Bank, S&L, Holding Co.
0.04%
0.04%
Average, All Service Industries
0.91%
0.80%
1993
1994
1995 Cost of Risk Survey - Tillinghast-Towers Perrin Organization

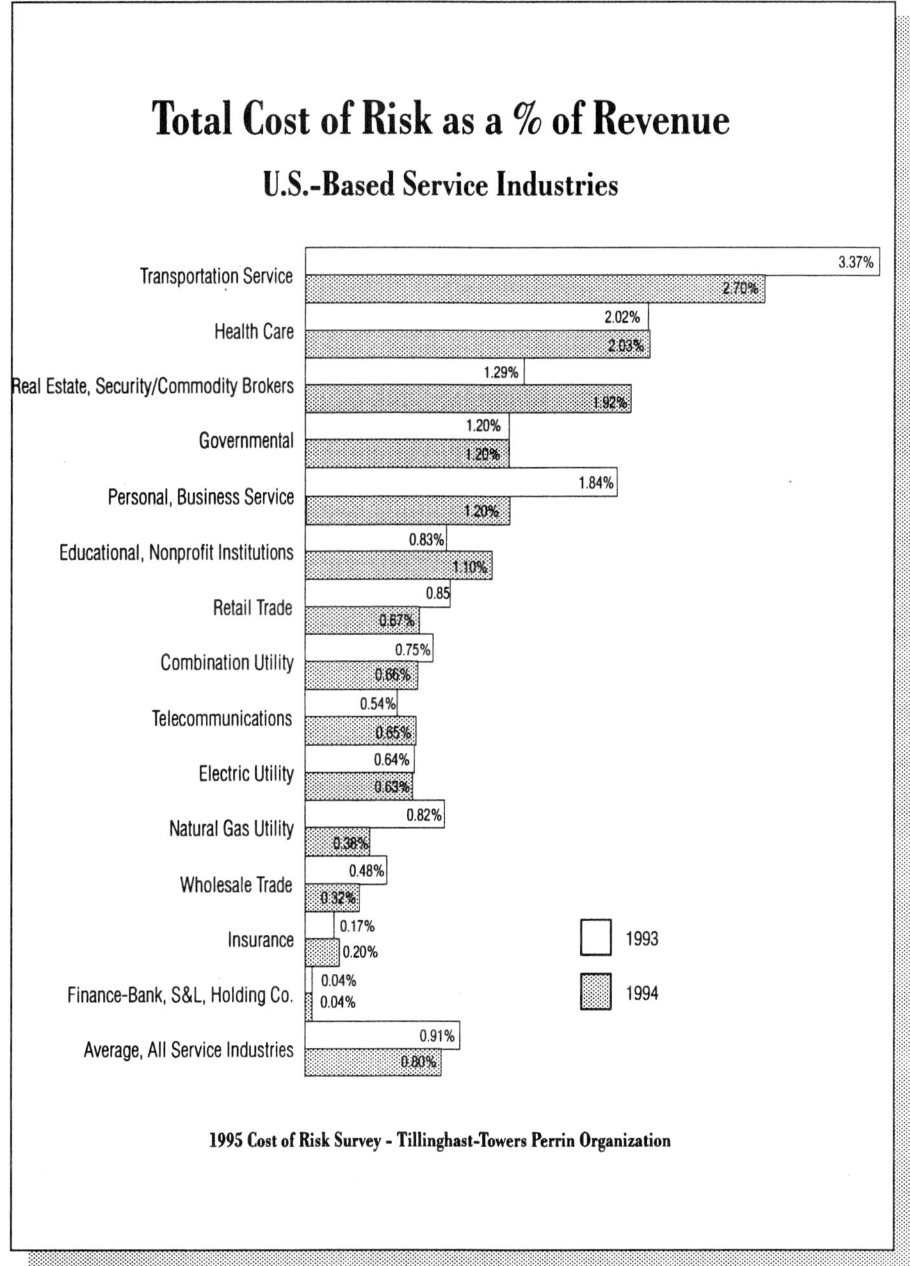

Figure 8-8

of the cost of risk has been excluded from the Cost of Risk Survey questionnaire since 1990."

We have delayed highlighting this point until now in order for readers to see the obvious depth of research needed to enable for two highly respected organizations to provide thousands of risk management leaders with this detailed report on costs.

We suggest that there could be several reasons for the cost of risk being in a difficult position to retrieve or differentiate from normal costs of doing business, as stated above, or as indicated by Dr. Veltri in his research on Accident Costs (Veltri 1988).

1. There are corporate leaders who truly recognize EHS as a moral, ethical responsibility of any well-managed organizations, and thus an imperative that accompanies accountability to making a profit. In fact, controlling the costs of risk adds substantially to the maximizing of profit.
2. There are those that, under their labor relations contract, have made EHS an integral part of every activity of their business.
3. There are those who believe EHS is one of the activities, in the operation of a business, that doesn't cost very much (less than 1% of revenue) and seems to fulfill some obligations in employee relations.
4. There are those, essentially motivated by profit and loss, who believe these activities cost very little and this buried cost is of inconsequential concern.
5. There are those, like their organizations founding fathers, that have not accepted the fact that accidents are caused and can be controlled, they are not just a standard part of the cost of doing business.

What appears to be a statistical dilemma may, in fact, be a blessing in disguise. A similar find was reported in an article, "An Accident Cost Impact Model: The Direct Cost Component" by Dr. Anthony Veltri, a professor at Oregon State University, in Vol. 21, 1990, issue of the *Journal of Safety Research* distributed by the National Safety Council.

Dr. Veltri points out that among findings from status studies in a variety of businesses, this interesting fact:

> Direct accident costs are frequently hidden in variable expense, portions of financial statements, and as a result are recognized as a necessary part of any producing and servicing activity (Veltri, 1988).

In the light of legislative requirements that include such laws as OSHA, EPA, RCRA, SARAIII, CAA, CWA TSCA and others, fines and penalties that now reach millions, liability that bankrupts giant organizations, prison sentences and a public whose negative opinion can stop production lines, there is no longer any question of EHS program existence. EHS today must be integrated into every facet of an organization. Risk control must become an important calculated normal cost of doing business for an organization with any hope of competing in today's business world.

We congratulate the NSC for their leadership example in the costing of accidents and suggest that the inclusion of appropriate coverage of these costs in the total cost of risk is more than substantiated by the studies discussed by us and the researchers referred to in NSC references. (See Appendix 2).

The pioneering work of Tillinghast-Towers Perrin and RIMS in their continuous improvement in the methodology for the costing of risk have also prepared the way for a simple updating of Barrows related standard, basically unchanged since 1962.

4. Benefits of an Integrated System Approach to Safety/EHS

Hundreds of top executives around the world like those quoted in Chapter 5 attest to the many benefits of their Safety/EHS program. The list shown here is a composite of them taken from specific communications we have had with these executives.

Benefits of Safety/EHS

- Reduced absenteeism
- Fewer grievances and gripes
- Fewer liability cases
- Smaller court awards
- Reduced insurance cost
- Compliance with safety and health legislation
- Control of injuries and illnesses
- Compliance with environmental legislation
- Preparation for natural disasters
- Preparation for general emergencies
- Fire and explosion prevention
- Property damage control
- Reduction of catastrophe potential
- Improved housekeeping
- Waste control
- Fewer employee errors
- Better quality control
- Continuous management improvement

- Better public image
- Fire and explosion prevention
- A "caring" culture
- Improved job pride
- Increased productivity
- Big potential cost reduction
- Profit and budget improvement

In the light of the evidence presented in this text, it is scarcely an exaggeration to say that an integrated system approach to safety/HSE will reduce injuries and illness, damage and defects, mistakes and waste, and by this action make a substantial improvement on the bottom line.

The direct Cost of Risk reported by more than 500 American companies in 1994 averaged .73% of their revenues (see page 297) or 12.7% of their net profit. We have also presented numerous historical cost relationship studies of direct to indirect costs. From Heinrich to the 1993 British HSE study they confirm that this relationship is at least 1:4. Conservatively applying a 1:2 ratio of direct to indirect, there is a high reliability in saying:

> The average direct and indirect Cost of Risk of American companies is a minimum of 25% of their net profit.

Controlling a company's Cost of Risk can result in immediate improvement in The Bottom Line.

Selected References

Bird, Frank E., Jr. and George L. Germain. *Practical Loss Control Leadership*. Loganville, GA: Institute Publishing, 1990.

Bird, Frank E., Jr. and George L. Germain. *Damage Control.* New York: American Management Association, 1966.

Business Week, *Corporate Scoreboard,* March 5, 1995

Grimaldi, John V. and Rollin H. Simonds. *Safety Management,* (Fifth Edition). New York: Irwin, 1994.

Health and Safety Executive, *The Costs of Accidents at Work.* Sheffield, UK: HSE, 1993.

Heinrich, H.W. *Industrial Accident Prevention* (First Edition). New York: McGraw-Hill Company, 1931.

National Safety Council, *Accident Facts* (1995 Edition). Itasca, IL: NSC, 1995.

Tillinghast-Towers Perrin, Risk and Insurance Management Society, 1995 Edition, *Cost of Risk Survey*, Tillinghast-Towers Perrin Risk management Publishers, New York, N.Y.

Simonds, Rollin H. and John V. Grimaldi; *Safety Management,* Revised Edition 1963, Richard D. Irwin Inc., Homewood, IL.

Veltri, Dr. Anthony, "An Accident Cost Impact Model: The Direct Cost Component." *Journal of Safety Research*, Vol. 21, 1990, National Safety Council.

1996 Cost of Risk Survey

A joint project of the Risk and Insurance Management Society, Inc. and Tillinghast-Towers Perrin

(Please print name exactly as you with it to appear, or attach business card.

Name:

Title:

Company:

Address:

City:

State/Province: Zip/Postal Code:

Telephone: Fax:

Please indicate your chapter affiliation below:
❑ RIMS Member/ Chapter ____________

❑ Non-RIMS Member Association Name: ____________

(e.g., AHRRM, PRIMA, URMIA)
Did your organization respond to this survey last year?
❑ Yes ❑ No

May we list your organization as a survey participant in the 1996 report? (Individual responses, of course, will remain confidential.) ❑ Yes ❑ No

If it is necessary to clarify any responses, whom should we contact? (If different from above)

Name: ______________________________

Phone: ______________________ Fax: ______________________

Please Note:

(1) WE'VE MADE CHANGES: To streamline this questionnaire and address your feedback, there have been several changes to response formats. PLEASE READ CAREFULLY

(2) Financial data requested in question 3.0 (actual or estimated) is mandatory or your questionnaire cannot be used. Tillinghast-Towers Perrin will preserve in confidence any company-specific information disclosed in connection with this survey.

(3) Each organization that completes the survey providing usable information by the deadline will receive a complimentary copy of the 1996 Cost of Risk Survey report. All others may purchase the report for $300.00

(4) Risk management pools/insurance trusts: If your organization operates a pool or trust on behalf of other entities do not fill out this questionnaire as your data will not be useful for the purposes of this survey.

Please return your completed survey by April 15, 1996 to:

1996 Cost of Risk Survey
Risk and Insurance Management Society, Inc., 655 Third Avenue, New York, New York 10017

If you have any questions while completing this survey, please contact Cecelia Cooper, RIMS, at (212) 286-9292, ext. 239 or Suzanne Greenlee, Tillinghast-Towers Perrin, at (703) 527-7500.

A RIMS Research Committee Endorsed Project

1996 Cost of Risk Survey

The 1995 reporting period for premiums, losses and outside service costs is the policy year ending between July 1, 1995 and June 30, 1996. It is desirable, but not imperative, that fiscal year data coincide precisely with insurance policy dates. For dollar amounts, show all significant figures (no symbols please), rounded to the nearest dollar, Loss data should reflect "ultimate expected" amounts for the 12 month reporting period only.

PART 1: DEMOGRAPHICS

1.0 Organization Domicile (circle one):

1. U.S. based
2. Canadian Based
3. Non U.S. or Canadian based

1.1. Responses are in: (circle one) 1. U.S. dollars 2. Canadian dollars

2.0 Industry

Please provide a primary and secondary 4-digit Standard Industrial Classification (SIC) code that is representative of your organization's primary activities as determined by the product, group of products produced or handled, or service rendered. (This information is available from your tax department. See the SIC code list at the end of the survey.)

A. primary SIC code________________ B. secondary SIC code: ________________

2.1 Please provide a one- to three-word description of your organization's primary industry or service, e.g., primary care hospital, commercial bank, clothing retailer, etc. ________________________

3.0 Worldwide Financial Data

The financial requested below is REQUIRED for survey completion. Cost figures cannot be included in the analysis without revenue to use as a base. Fiscal year data reported should be fore the year ending between July 1, 1995 and June 30, 1996. If actual data cannot be supplied, please provide best estimates.

		U.S. Operations	Canadian Operations	Non U.S./ Canadian Operations
3.1	Gross revenue/sales	$______	$______	$______
3.2	Operating budget (if public entity, non-profit organization, university, etc.)	$______	$______	$______
3.3	Deposits (if a banking organization)	$______	$______	$______
3.4	Year-end number of FTE employees (full-time equivalent)	$______	$______	$______
3.5	Payroll (salaries and wages only)	$______	$______	$______
3.6	Year-end total assets (from balance sheet)	$______	$______	$______

PART II: ADMINISTRATIVE INFORMATION

4.0 Risk Management/Insurance Staffing

4.1 In which department is the person responsible for risk/insurance duties located? (circle one)

1. Risk Management/Insurance
2. Human Resources
3. Finance/Treasury
4. Safety/Loss Control
5. Legal
6. Other (please specify): ____________________

4.2 Function to which the top risk management professional reported in 1995: (circle one)

1. Treasury
2. Legal
3. Finance/CFO
4. President/CEO/Governor/Mayor
5. Administration
6. Human Resources
7. Other (please specify): ____________________

4.3 Number of full-time equivalent employees with risk/insurance duties (e.g., claims, insured litigation, loss control) at year end 1995. Include both full-time employees and fractional employees who have shared duties.

Department	Professional	Clerical
A. Risk Management/Insurance	____________	____________
B. Human Resources	____________	____________
C. Legal	____________	____________
D. Safety/Loss Control	____________	____________
E. Other: ________	____________	____________

4.4. How has the number of risk management/insurance department employees changed in the last two years?

1 Increased 2 Decreased 3 No change

4.5. If the number of risk management/insurance department employees has decreased in the last two years, please indicate reasons for decrease (circle all that apply):

1 Outsourcing 2 Downsizing 3 Technology 4 Reduced functions

5 Other (please specify): ______________________________

4.6 Do you anticipate any change in the number of persons engaged in risk management activities in the next two years?

1 Increase 2 Decrease 3 No change

4.7 If the number of risk management/insurance department employees is anticipated to decrease in the next two years, please indicate reasons for decrease (circle all that apply):

1 Outsourcing 2 Downsizing 3 Technology 4 Reduced functions

5 Other (please specify): ______________________________

4.8 Where was the primary responsibility for the following functions in 1995? (circle only one each line)

	Function	Risk Mgmt. Dept.	Human Resources	Legal	Third Party Administrator	Other (Specify)
Insurance Purchase	Property/Liability	1	2	3	4	5.___
	Workers Comp.	1	2	3	4	5.___
	Health/Medical	1	2	3	4	5.___
Claims Management	Property	1	2	3	4	5.___
	Liability	1	2	3	4	5.___
	Workers Comp.	1	2	3	4	5.___
	Health/Medical	1	2	3	4	5.___
	Environmental	1	2	3	4	5.___
Risk Control/Engineering	Property	1	2	3	4	5.___
	Liability	1	2	3	4	5.___
	Workers Comp	1	2	3	4	5.___
	Environmental	1	2	3	4	5.___
	Security	1	2	3	4	5.___
	Disaster/Catastrophe Planning	1	2	3	4	5.___

5.0 1995 Risk Management/Insurance Administrative Budget

Include only departmental expenses and overhead burdens for only those staff with risk/insurance related duties. DO NOT INCLUDE premiums, losses or service provider costs which are reported in other parts of this survey.

Department	1995 Budget (actual if available)
A. Risk Management/Insurance	$________________
B. Human Resources	$________________
C. Legal	$________________
D. Safety/Loss Control	$________________
E. Other ____________	$________________

Part III: WORKERS COMPENSATION RISK FINANCING PROGRAM

6.0 Risk Financing Structure

How did you finance workers compensation in 1995? (Circle all methods used)

1. Fixed-cost insurance
2. Retrospectively rated insurance
3. Deductible insurance plan
4. Dividend insurance plan
5. Qualified self-insurance
6. Association/group program
7. Provincial government
8. Other: ____________________

7.0 Workers Compensation Premiums, Retention Levels, and Retained Losses

This section is for workers compensation and employer's liability costs and claims, whether or not reported or paid, which actually occurred in the 1995 reporting period. For purposes of this questionnaire, wholly owned captives are not third-party insurers.

Premium includes all fixed costs paid to third parties such as guaranteed cost policies, minimum premiums for loss-responsive plans, fronting charges, letter of credit charges, assessments, excess WC premiums, agent commissions, captive reinsurance, and taxes on these amounts.

Retention level is the maximum per-occurrence loss limitation, whether the program is retrospective, fronted, deductible or self-insurance. If reinsured to a wholly owned captive, show the captive retention.

Retained losses includes the "ultimate expected" amount of all variable (loss driven) costs such as paid losses, case reserves, IBNR estimates, allocated loss adjustment expenses (ALAE), claim handling (loss conversion) charges, and taxes on these amounts. **Losses retained by a captive should be reported here.**

	Workers Compensation/ Employer's Liability	Premium (Fixed Costs)	Retention Level	Retained Losses (Variable Costs)
7.1	A. U.S. domestic programs including state funds, assigned risk pools, USL&H,FELA, Jones Act	$________	$________	$________
	B. Canadian Provincial Workers Compensation Boards	$________	$________	$________
	C Non U.S./Canadian programs	$________	$________	$________
	D Other________	$________	$________	$________
	Total	$________		$________

Note: if a captive is used, please split the captive premium between fixed costs and variable costs, to the best of your knowledge.

7.2. Do you use actuarial analysis to estimate ultimate losses?

1 Yes 2 No

PART IV: LIABILITY RISK FINANCING PROGRAM

8.0 Liability Premiums, Limits, Retention Levels, and Retained Losses

This section is for all liability costs for claims relating to the 1995 reporting period whether or not reported or paid. For purposes of this questionnaire, wholly owned captives are not third-party insurers.

Premium includes all fixed costs paid to third parties such as guaranteed cost policies, minimum premiums for loss-responsive plans, fronting charges, letter of credit charges, assessments, excess WC premiums, agent commissions, captive reinsurance, and taxes on these amounts.

Retention level is the maximum per-occurrence loss limitation, whether the program is retrospective, fronted, deductible or self-insurance. If reinsured to a wholly owned captive, show the captive retention.

Retained losses includes the "ultimate expected" amount of all variable (loss driven) costs such as paid losses, case reserves, IBNR estimates, allocated loss adjustment expenses (ALAE), claim handling (loss conversion) charges, and taxes on these amounts. **Losses retained by a captive should be reported here.**

For each line of coverage carried, be sure to fill in all columns. If a line of coverage is not carried, or is included elsewhere, leave blank.

	LINE OF COVERAGE	PREMIUM (FIXED COSTS)	POLICY LIMITS	RETENTION LEVELS	RETAINED LOSSES (VARIABLE COSTS)
8.1	Primary general liability	$______	$______	$______	$______
8.2	Primary auto liability	$______	$______	$______	$______
8.3	Primary foreign G.L./A.L.	$______	$______	$______	$______
8.4	Products liability	$______	$______	$______	
8.5	Umbrella/excess liability	$______	$______		$______
8.6	Professional liability	$______	$______	$______	$______
8.7	D&O liability	$______	$______	$______	$______
8.8	Pollution/environmental liability	$______	$______	$______	$______
8.9	Fiduciary liability	$______	$______	$______	$______
8.10	Aviation liability	$______	$______	$______	$______
8.11	Marine/P&I/ship repair liability	$______	$______	$______	$______
8.12	Nuclear liability	$______	$______	$______	$______
8.13	Other liability: ________	$______	$______	$______	$______
8.14	Other liability: ________	$______	$______	$______	$______
	Total	$______			$______

Note: if a captive is used, please split the captive premium between fixed costs and variable costs, to the best of your knowledge.

8.15 Do you use actuarial analysis to estimate ultimate losses?

1 Yes 2 No

PART V: PROPERTY RISK FINANCING PROGRAM

9.0 Property Premiums, Insured Values, Deductibles, and Retained Losses

This section is for property and other physical damage costs for claims occurring in the 1995 reporting period. For purses of this questionnaire, wholly owned captives are not third-party insurers.

Premium includes all fixed costs paid to third parties such as guaranteed cost policies, minimum premiums for loss-responsive plans, assessments, agent commissions, captive reinsurance, and taxes on these amounts.

Insured value is the total amount insured under the policy, or the limits/sublimits available in the policy.

Deductible level is the maximum per-loss retention, including captive retentions. For policies with multiple deductibles, please report the "most characteristic" deductible.

Retained losses includes all variable (loss driven) costs such as paid and reserved amounts, claim handling charges, and taxes. **Losses retained by a captive should be included here.**

For each line of coverage carried, be sure to fill in all columns. If a line of coverage is not carried, or is included elsewhere, leave blank.

	LINE OF COVERAGE	PREMIUM (FIXED COSTS)	INSURED VALUE OR LIMIT	DEDUCTIBLE LEVEL	RETAINED LOSSES VARIABLE COSTS)
9.1	Direct damage (all risk)	$______	$______	$______	$______
9.2	Business interruption/extra expense	$______	$______	$______	$______
9.3	Boiler & machinery	$______	$______	$______	$______
9.4	Auto/truck physical damage	$______	$______	$______	$______
9.5	Flood	$______	$______	$______	$______
9.6	Earthquake (except CA)	$______	$______	$______	$______
9.7	California earthquake	$______	$______	$______	$______
9.8	Fidelity/crime (including financial institution blanket bond)	$______	$______	$______	$______
9.9	Inland marine/property floaters	$______	$______	$______	$______
9.10	Marine hull & machinery	$______	$______	$______	$______
9.11	Cargo	$______	$______	$______	$______
9.12	Control of well/OEE	$______	$______	$______	$______
9.13	Foreign property (if not included above)	$______	$______	$______	$______
9.14	Aircraft hull	$______	$______	$______	$______
9.15	Builders risk	$______	$______	$______	$______
9.16	Other property: ______	$______	$______	$______	$______
9.17	Other property: ______	$______	$______	$______	$______
	Total	$______			$______

Note: if a captive is used, please split the captive premium between fixed costs and variable costs, to the best of your knowledge.

PART VI: OUTSIDE SERVICES

10.0 Insurance Agents/Brokers

This section refers primarily to retail agent/broker relationships. Do not include wholesale intermediaries UNLESS they were engaged and paid directly by your organization or captive. For example, a London or surplus lines broker used by your retail agent should not be reflected here.

10.1 How many insurance agencies/brokerage firms did your organization use in 1995? ___________

10.2 How were your agents/brokers compensated? (circle one)

1 Straight commission

2 Fee for service (includes negotiated commissions)

3 Combination of both

Note: If you paid premiums to a captive which in turn paid an agent/broker a negotiated commission or fee, report his as a fee for service.

10.3 For those lines of coverage where agents/brokers were compensated on a **straight commission basis**, what was the commission rate as a percentage of premiums? (Lease blank if you pay only a fee for service.)

COVERAGE	% OF PREMIUM	COVERAGE	% OF PREMIUM
A. Property/Business Interruption	____	E. D&O Liability	____
B. Primary General Liability	____	F. Environmental Impairment	____
C. Primary Auto Liability	____	G. Umbrella/Excess Liability	____
D. Professional Liability	____	H. Worker's Compensation	____

10.4 What was the total amount of compensation paid to agents/brokers?

As commission $ _________ As fees $ _________

11.0 Other Service Providers

This section refers to service provider costs **other than** those included in the premiums, retained losses, or agent/broker fees.

A. Outside consulting/actuarial fees $ _________

B. Risk management information systems $ _________

C. Captive management costs $ _________

D. Miscellaneous other costs for outside services $ _________

Note: Where these costs are paid from premiums to a captive, if possible, please report appropriate amounts here and subtract from premiums.

12.0	**Captives**		
	A. Workers Compensation	1 Yes	2 No
	B. Liability	1 Yes	2 No
	C. Property	1 Yes	2 No

13.0 Risk Management Information Systems

13.1 What type of computer record keeping system do you use? (circle all that apply)

1 None	4 Timeshare vendor
2 In-house mainframe	5 Broker/insurer
3 In-house PC./mini	6 Other: ________

13.2 What key tasks does your system(s) perform? (circle all that apply)

1 Claims management	4 Tracking ins. certificates
2 Loss analysis & forecasting	5 Tracking property value
3 Policy management	6 Other: ________

PART VII: Additional Information

Additional comments/suggestions for future surveys

____

____

____

____

____

____

Thank you for your participation. Please return this form by April 15, 1996 to:

1996 Cost of Risk Survey
Risk and Insurance Management Society, Inc.
655 Third Avenue
New York, New York 10017

Note: The Cost of Risk Survey form is in the appendix of this book with permission of the survey publishers.

Standard Industrial Classification (SIC) Codes

01 AGRICULTURAL PRODUCTION - CROPS
0110 Cash Grains
0130 Field Crops, Except Cash Grains
0160 Vegetables and Melons
1070 Fruits and Tree Nuts
1080 Horticultural Specialties
1090 General Farms, Primarily Crop

02 AGRICULTURAL PRODUCTION - LIVESTOCK
0210 Livestock, except Dairy and Poultry
0250 Poultry and Eggs
0270 Animal Specialties
0290 General Farms, Primarily Livestock and Animal Specialties

07 AGRICULTURAL SERVICES
0710 Soil Preparation Services
0720 Crop Services
0740 Veterinary Services
0750 Animal Services, Except Veterinary
0760 Farm Labor and Management Services
0780 Landscape and Horticultural Services

08 FORESTRY
0810 Timber Tracts
0830 Forest Nurseries and Gathering of Forest Products
0850 Forestry Services

09 FISHING, HUNTING, AND TRAPPING
0910 Commercial Fishing
0920 fish Hatcheries and Preserves
0970 Hunting and Trapping, and Game Propagation

10 METAL MINING
1010 Iron Ores
1020 Copper Ores
1030 Lead and Zinc Ores
1040 Gold and Silver Ores
1060 Ferroalloy Ores, Except Vanadium
1080 Metal Mining Services
1090 Miscellaneous Metal Ores

12 COAL MINING
1220 Bituminous Coal and Lignite Mining
1230 Anthracite Mining
1240 Coal Mining Services

13 OIL AND GAS EXTRACTION
1310 Crude Petroleum and Natural Gas
1320 Natural Gas Liquids
1380 Oil and Gas Field Services

14 NONMETALLIC MINERALS, EXCEPT FUELS
1410 Dimension Stone
1420 Crushed and Broken Stone, Including Riprap
1440 Sand and Gravel
1450 Clay, Ceramic and Refractory Minerals
1470 Chemical and Fertilizer Mineral Mining
1480 Nonmetallic Mineral Services, Except Fuels
1490 Miscellaneous Nonmetallic Minerals, Except Fuels

15 GENERAL BUILDING CONTRACTORS
1520 General Building Contractors - Residential Buildings
1530 Operative Builders
1540 General Building Contractors - Nonresidential Buildings

16 HEAVY CONSTRUCTION, EXCEPT BUILDING
1610 Highway and Street Construction, Except Elevated Highways
1620 Heavy Construction, Except Highway and Street Construction

17 SPECIAL TRADE CONTRACTORS
1710 Plumbing, Heating and Air-Conditioning
1720 Painting and Paper Hanging
1730 Electrical Work
1740 Masonry, Stonework, Tile Setting, and Plastering
1750 Carpentry and Floor Work
1760 Roofing, Siding, and Sheet Metal Work
1770 Concrete Work
1780 Water Well Drilling
1790 Miscellaneous Special Trade Contractors

20 MANUFACTURING: FOOD AND KINDRED PRODUCTS
2010 Meat Products
2020 Dairy Products
2030 Canned, Frozen, and Preserved Fruits, Vegetables, and Food Specialties
2040 Grain Mill Products
2050 Bakery Products
2060 Sugar and Confectionery Products
2070 Fats and Oils
2080 Beverages
2090 Miscellaneous Food Preparations and Kindred Products

21 MANUFACTURING: TOBACCO PRODUCTS
2110 Cigarettes
2120 Cigars
2130 Chewing and Smoking Tobacco
2140 Tobacco Stemming and Redrying

22 MANUFACTURING: TEXTILE MILL PRODUCTS
2210 Broadwoven Fabric Mills, Cotton
2220 Broadwoven Fabric Mills, Manmade
2230 Broadwoven Fabric Mills, Wool
2240 Narrow Fabric Mills
2250 Knitting Mills
2260 Textile Finishing, Except Wool
2270 Carpets and Rugs
2280 Yard and Tread Mills
2290 Miscellaneous Textile Goods

23 MANUFACTURING: APPAREL AND OTHER TEXTILE PRODUCTS
2310 Men's and Boy's Suits and Coats
2320 Men's and Boy's Furnishings
2330 Women's and Misses' Outerwear
2340 Women's and Misses' Undergarments
2350 Hats, Caps, and Millinery
2360 Girl's and Children's Outerwear
2370 Fur Goods
2380 Miscellaneous Apparel and Accessories
2390 Miscellaneous Fabricated Textile Products

24 MANUFACTURING: LUMBER AND WOOD PRODUCTS
2410 Logging
2420 Sawmills and Planing Mills
2430 Millwork, Plywood & Structural Members
2440 Wood Containers
2450 Wood Buildings and Mobil Homes

25 MANUFACTURING: FURNITURE AND FIXTURES
2510 Household Furniture
2520 Office Furniture
2530 Public Building and Related Furniture
2540 Partitions and Fixtures
2590 Miscellaneous Furniture and Fixtures

26 MANUFACTURING: PAPER AND ALLIED PRODUCTS
2610 Pulp Mills
2620 Paper Mills
2630 Paperboard Mills
2650 Paperboard Containers and Boxes
2670 Miscellaneous Converted Paper Products

27 PRINTING AND PUBLISHING
2710 Newspapers
2720 Periodicals
2730 Books
2740 Miscellaneous Publishing
2750 Commercial Publishing
2760 Manifold Business Forms
2770 Greeting Cards
2780 Blankbooks and Bookbinding
2790 Printing Trade Services

28 MANUFACTURING: CHEMICALS AND ALLIED PRODUCTS
2810 Industrial Inorganic Chemicals
2820 Plastics Materials and Synthetics
2830 Drugs
2840 Soap, Cleaners, and Toilet Goods
2850 Paints and Allied Products
2860 Industrial Organic Chemicals
2870 Agricultural Chemicals
2890 Miscellaneous Chemical Products

29 MANUFACTURING: PETROLEUM AND COAL PRODUCTS
2910 Petroleum Refining
2950 Asphalt Paving and Roofing Materials
2990 Miscellaneous Petroleum and Coal Products

30 MANUFACTURING: RUBBER AND MISC. PLASTICS PRODUCTS
3010 Tires and Inner Tubes
3020 Rubber and Plastics Footwear
3050 Hose and Belting and Gaskets and Packing
3060 Fabricated Rubber Products, NEC
3080 Miscellaneous Plastics Products, NEC

31 MANUFACTURING: LEATHER AND LEATHER PRODUCTS
3110 Leather Tanning and Finishing
3130 Footwear Cut Stock
3140 Footwear, Except Rubber
3150 Leather Gloves and Mittens
3160 Luggage
3170 Handbags and Personal Leather Goods
3190 Leather Goods, NEC

32 MANUFACTURING: STONE, CLAY AND GLASS PRODUCTS
3210 Flat Glass
3220 Glass and Glassware, Pressed or Blown
3230 Products or Purchased Glass
3240 Cement, Hydraulic
3250 Structural Clay Products
3260 Pottery and Related Products
3270 Concrete, Gypsum, and Plaster Products
3280 Cut Stone and Stone Products
3290 Miscellaneous Nonmetallic Mineral Products

33 MANUFACTURING: PRIMARY METAL INDUSTRIES
3310 Blast Furnace and Basic Steel Products
3320 Iron and Steel Foundries
3330 Primary Nonferrous Metals
3340 Secondary Nonferrous Metals
3350 Nonferrous Rolling and Drawing
3360 Nonferrous Foundries (Casting)
3390 Miscellaneous Primary Metal Products

34 MANUFACTURING: FABRICATED METAL PRODUCTS
3410 Metal Cans and Shipping Containers
3420 Cutlery, Handtools, and Hardware
3430 Plumbing and Heating, Except Electric
3440 Fabricated Structural Metal Products
3450 Screw Machine Products, Bolts, etc.
3460 Metal Forgings and Stampings
3470 Metal Services, NEC
3480 Ordnance and Accessories, NEC
3490 Miscellaneous Fabricated Metal Products

35 MANUFACTURING: INDUSTRIAL & MACHINERY EQUIPMENT
3510 Engines and Turbines
3520 Farm and Garden Machinery
3530 Construction and Related Machinery
3540 Metalworking Machinery
3550 Special Industry Machinery
3560 General Industrial Machinery
3570 Computer and Office Equipment
3580 Refrigeration and Service Machinery
3590 Industrial Machinery, NEC

36 MANUFACTURING: ELECTRONIC & OTHER ELECTRIC EQUIPMENT
3610 Electric Distribution Equipment
3620 Electrical Industrial Apparatus
3630 Household Appliances
3640 Electric Lighting and Wiring Equipment
3650 Household Audio and Video Equipment
3660 Communications Equipment
3670 Electronic Components and Accessories
3690 Miscellaneous Electrical Equipment and Supplies

37 MANUFACTURING: TRANSPORTATION EQUIPMENT
3710 Motor Vehicles and Equipment
3720 Aircraft and Parts
3730 Ship and Boat Building and Repairing
3740 Railroad Equipment
3750 Motorcycles, Bicycles, and Parts
3760 Guided Missiles, Space Vehicles, Parts
3790 Miscellaneous Transportation Equipment

38 MANUFACTURING: INSTRUMENTS AND RELATED PRODUCTS
3810 Search and Navigation Equipment
3820 Measuring and Controlling Devices
3840 Medical Instruments and Supplies
3850 Ophthalmic Goods
3860 Photographic Equipment and Supplies
3870 Watches, Clocks, Watchcases and Parts

39 MISCELLANEOUS MANUFACTURING INDUSTRIES
3910 Jewelry, Silverware, and Plated Ware
3930 Musical Instruments
3940 toys and Sporting Goods
3950 Pens, Pencils, Office, and Art Supplies
3990 Miscellaneous Manufactures

40 RAILROAD TRANSPORTATION
4010 Railroads

41 LOCAL AND INTERURBAN PASSENGER TRANSIT
4110 Local and Suburban Transportation
4120 Taxicabs
4130 Intercity and Rural Bus Transportation
4140 Bus Charter Service
4150 School Buses
4170 Bus Terminal and Service Facilities

42 TRUCKING AND WAREHOUSING
4210 Trucking and Courier Services, Except Air
4220 Public Warehousing and Storage
4230 Trucking Terminal Facilities

43 U.S. POSTAL SERVICE
4310 U.S. Postal Service

44 WATER TRANSPORTATION
4410 Deep Sea Foreign Transportation of Freight
4420 Deep Sea Domestic Transportation of Freight
4430 Freight Transportation on the Great Lakes
4440 Water Transportation of Freight, NEC
4480 Water Transportation of Passengers
4490 Water Transportation Services

45 TRANSPORTATION BY AIR
4510 Air Transportation, Scheduled
4520 Air Transportation, Nonscheduled
4580 Airports, Flying Fields, and Services

46 PIPELINES, EXCEPT NATURAL GAS
4610 Pipelines, Except Natural Gas

47 TRANSPORTATION SERVICES
4720 Passenger Transportation Arrangement
4730 Freight Transportation Arrangement
4740 Rental of Railroad Cars
4780 Miscellaneous Transportation Services

48 COMMUNICATIONS
4810 Telephone Communications
4820 Telegraph and Other Communications
4830 Radio and Television Broadcasting
4840 Cable and Other Pay TV Services
4890 communications Services, NEC

49 ELECTRIC, GAS, AND SANITARY SERVICES
4910 Electric Services
4920 Gas Production and Distribution
4930 Combination Utility Services
4940 Water Supply
4950 Sanitary Services
4960 Steam and Air-Conditioning Supply
4970 Irrigation Systems

50 WHOLESALE TRADE - DURABLE GOODS
5010 Motor Vehicles, Parts and Supplies
5020 Furniture and Homefurnishings
5030 Lumber and Construction Materials
5040 Professional and Commercial Equipment
5050 Metals and Minerals, Except Petroleum
5060 Electrical Goods
5070 Hardware, Plumbing and Heating Equipment
5080 Machinery, Equipment and Supplies
5090 Miscellaneous Durable Goods

51 WHOLESALE TRADE - NONDURABLE GOODS
5110 Paper and Paper Products
5120 Drugs, Proprietaries and Sundries
5130 Apparel, Piece Goods, and Notions
5140 Groceries and Related Products
5150 Farm-Product Raw Materials
5160 Chemicals and Allied Products
5170 Petroleum and Petroleum Products
5180 Beer, Wine and Distilled Products
5190 Miscellaneous Nondurable Goods

52 RETAIL TRADE - BUILDING MATERIALS & GARDEN SUPPLIES
5210 Lumber and Other Building Materials
5230 Paint, Glass, and Wallpaper Stores
5350 Hardware Stores
5260 Retail Nurseries and Garden Stores
5270 Mobile Home Dealers

53 GENERAL MERCHANDISE STORES
5310 Department Stores
5330 Variety Stores
5390 Miscellaneous General Merchandise Stores
54 FOOD STORES
5410 Grocery Stores
5420 Meat and Fish Markets
5430 Fruit and Vegetable Markets
5440 Candy, Nut, and Confectionery Stores
5450 Dairy Products Stores
5460 Retail Bakeries
5490 Miscellaneous Food Stores
55 AUTOMOTIVE DEALERS & SERVICE STATIONS
5510 New and Used Car Dealers
5520 Used Car Dealers
5530 Auto and Home Supply Stores
5540 Gasoline Service Stations
5550 Boat Dealers
5560 Recreational Vehicle Dealers
5570 Motorcycle Dealers
5590 Automotive Dealers, NEC
56 APPAREL AND ACCESSORY STORES
5610 Men's and Boy's Clothing Stores
5620 Women's Clothing Stores
5630 Women's Accessory And Specialty Stores
5640 Children;s and Infant's Wear Stores
5650 Family Clothing Stores
5660 Shoe Stores
5690 Miscellaneous Apparel and Accessory Stores
57 FURNITURE AND HOMEFURNISHING STORES
5710 Furniture and Homefurnishing Stores
5720 Household Appliance Stores
5730 Radio, Television and Computer Stores
58 EATING AND DRINKING PLACES
5810 Eating and Drinking Places
59 MISCELLANEOUS RETAIL
5910 Drug Stores and Proprietary Stores
5920 Liquor Stores
5930 Used Merchandise Stores
5940 Miscellaneous Shopping Goods Stores
5960 Nonstore Retailers
5980 Fuel Dealers
5990 Retail Stores, NEC
60 DEPOSITORY INSTITUTIONS
6010 Central Reserve Depositories
6020 Commercial Banks
6030 Savings Institutions
6060 Credit Unions
6080 Foreign Banks and Branches and Agencies
6090 Functions Closely Related to Banking
61 NONDEPOSITORY INSTITUTIONS
6110 Federal and Federally-Sponsored Credit
6140 Personal Credit Institutions
6150 Business Credit Institutions
6160 Mortgage Bankers and Brokers
62 SECURITY AND COMMODITY BROKERS
6210 Security Brokers and Dealers
6220 Commodity Contracts Brokers, Dealers
6230 Security and Commodity Exchanges
6280 Security and Commodity Services
63 INSURANCE CARRIERS
6310 Life Insurance
6320 Medical Service and Health Insurance
6330 Fire, marine and Casualty Insurance
6350 Surety Insurance
6360 Title Insurance
6370 Pension, Health, and Welfare Funds
6390 Insurance Carriers, NEC
64 INSURANCE AGENTS, BROKERS AND SERVICE
6410 Insurance Agents, Brokers and Service
65 REAL ESTATE
6510 Real Estate Operators and Lessors
6530 Real Estate Agents and Managers
6540 Title Abstract Offices
6550 Subdividers and Developers
67 HOLDING AND OTHER INVESTMENT OFFICES
6710 Holding Offices
6720 Investment Offices
6730 Trusts
6790 Miscellaneous Investing
70 HOTELS AND OTHER LODGING PLACES
7010 Hotels and Motels
7020 Rooming and Boarding Houses
7030 Camps and Recreational Vehicle Parks
7040 Membership-Basis Organization Hotels
72 PERSONAL SERVICES
7210 Laundry, Cleaning, and Garment Services
7220 Photographic Studios, Portrait
7230 Beauty Shops
7240 Barber Shops
7250 Shoe Repair and Shoeshine Parlors
7260 Funeral Service and Crematories
7290 Miscellaneous Personal Services
73 BUSINESS SERVICES
7310 Advertising
7320 Credit Reporting and Collection
7330 Mailing, Reproduction, Stenographic
7340 Services to Buildings
7350 Miscellaneous Equipment Rental and Leasing
7360 Personnel Supply Services
7370 Computer and Data Processing Services
7380 Miscellaneous Business Services
75 AUTO REPAIR, SERVICES AND PARKING
7510 Automotive Rentals, No Drivers
7520 Automobile Parking
7530 Automotive Repair Shops
7540 Automotive Services, Except Repair
76 MISCELLANEOUS REPAIR SERVICES
7620 Electrical Repair Shops
7630 Watch, Clock and Jewelry Repair
7640 Reupholstery and Furniture Repair
7690 Miscellaneous Repair Shops
78 MOTION PICTURES
7810 Motion Picture Production and Services
7820 Motion Picture Distribution Services
7830 Motion Picture Theaters
7840 Video Tape Rental
79 AMUSEMENT AND RECREATION SERVICES
7910 Dance Studios, Schools, and Halls
7920 Producers, Orchestras, Entertainers
7930 Bowling Centers
7940 Commercial Sports
7990 Miscellaneous Amusement, Recreation Services
80 HEALTH SERVICES
8010 Offices and Clinics of Medical Doctors
8020 Offices and Clinics of Dentists
8030 Offices of Osteopathic Physicians
8040 Offices of Other Health Practitioners
8050 Nursing and Personal Care Facilities
8060 Hospitals
8070 Medical and Dental Laboratories
8080 Home Health Care Practitioners
8090 Health and Allied Services, NEC
81 LEGAL SERVICES
8110 Legal Services
82 EDUCATIONAL SERVICES
8210 Elementary and Secondary Schools
8220 Colleges and Universities
8230 Libraries
8240 Vocational Schools
8290 Schools and Educational Services, NEC
83 SOCIAL SERVICES
8320 Individual and Family Services
8330 Job Training and Related Services
8350 Child Day Care Services
8360 Residential Care
8390 Social Services, NEC
84 MUSEUMS, BOTANICAL, ZOOLOGICAL GARDENS
8410 Museums and Art Galleries
8420 Botanical and Zoological Gardens
86 MEMBERSHIP ORGANIZATIONS
8610 Business Associations
8620 Professional Associations
8630 Labor Organizations
8640 Civic and Social Associations
8650 Political Organizations
8660 Religious Organizations
8690 Membership Organizations, NEC
87 ENGINEERING AND MANAGEMENT SERVICES
8710 Engineering and Architectural Services
8720 Accounting, Auditing and Bookkeeping
8730 Research and Testing Services
8740 Management and Public Relations

88 PRIVATE HOUSEHOLDS
8810 Private Households
89 GOVERNMENTAL ENTITIES
8940 Municipalities*
8950 Cities*
8960 Counties*
8970 States*
8980 Provinces*
8990 Services, NEC

NEC = Not Elsewhere Classified
*These codes have been created by RIMS.

Source:
Standard Industrial Classification Manual 1987

National Safety Council References and Methodology for Work Deaths and Injury Costs.

U.S. Department of Commerce. Resident population is used for computing rates.

Costs (pp.4-7). The procedures for estimating the economic losses due to fatal and nonfatal unintentional injuries were extensively revised for the 1993 edition of *Accident Facts*. New components were added, new benchmarks adopted, and a new discount rate assumed. All of these changes resulted in significantly higher cost estimates. For this reason, it must be re-emphasized that the cost estimates should not be compared to those in earlier editions of *Accident Facts.*

The Council's general philosophy underlying its cost estimates is that the figures represent income not received or expenses incurred because of fatal and nonfatal unintentional injuries. Stated this way, the Council's cost estimates are a measure of the economic impact of unintentional injuries and may be compared to other economic measures such as gross domestic product, per capital income, or personal consumption expenditures. (See page 83 and "lost quality of life" [in this section] for a discussion of injury costs for cost-benefit analysis.)

The general approach followed was to identify a benchmark unit cost for each component, adjust the benchmark to the current year using an appropriate inflator, estimate the number of cases to which the component applied, and compute the product. Where possible, benchmarks were obtained for each class: Motor Vehicle, Work, Home, and Public. The complete procedures are very complicated. It is possible here to give only a brief overview of the procedures. More detailed information is available on request from the Statistics Department.

Source: National Safety Council, *Accident Facts*, 1995.

Wage and productivity losses include the value of wages, fringe benefits, and household production for all classes, and travel delay for the Motor Vehicle class.

For fatalities, the present value of after-tax wages, fringe benefits, and household production was computed using the human capital method. The procedure incorporates data on life expectancy from the NCHS life tables, employment likelihood from the Bureau of Labor Statistics household survey, and mean earnings from the Bureau of the Census money income survey. The discount rate used was 4%, reduced from 6% used in earlier years. The present value obtained is highly sensitive to the discount rate; the lower the rate, the greater the present value.

For permanent partial disabilities, an average of 17% of earning power is lost (Berkowitz & Burton, 1987). The incident of permanent disabilities, adjusted to remove intentional injuries, was computed from data on hospitalized cases from the National Hospital Discharge Survey (NHDS) and non-hospitalized cases from the national Health Interview Survey and National Council on Compensation Insurance data on probabilities of disability by nature of injury and part of body injured.

For temporary disabilities, an average daily wage, fringe benefit, and household production loss was calculated and this was multiplied by the number of days of restricted activity from the NHIS.

Travel delay costs were obtained from the Council's estimates of the number of fatal, injury, and property damage crashes and an average delay cost per crash from Miller and others (1991).

Medical expenses, including ambulance and helicopter transport costs, were estimated for fatalities, hospitalized cases, and nonhospitalized cases in each class.

Source: National Safety Council, *Accident Facts*, 1995.

The incidence of hospitalized cases was derived from the NHDS data adjusted to eliminate intentional injuries. Average length of stay was benchmarked from Miller, Pindus, Douglass and Rossman (1993b) and adjusted to estimate lifetime length of stay. The cost per hospital day was benchmarked to the National Medical Expenditure Survey (NMES).

Nonhospitalized cases were estimated by taking the difference between total NHIS injuries and hospitalized cases. Average cost per case was based on NMES data adjusted for inflation and lifetime costs.

Medical cost of fatalities was benchmarked to data from the National Council on Compensation Insurance (1989) to which was added to cost of a premature funeral and coroner costs (Miller et al. 1991).

Cost per ambulance transport were benchmarked to NMES data and cost per helicopter transport were benchmarked to data in Miller and others (1993a). The number of cases transported was based on data from Rice and MacKenzie (1989) and the National Electronic Injury Surveillance System.

Administrative expenses include the administrative cost of private and public insurance, which represents the cost of having insurance, and police and legal costs.

The administrative cost of motor-vehicle insurance was the difference between premiums earned (adjusted to remove fire, theft, and casualty premiums) and pure losses incurred, based on data from A. M. Best. Workers' compensation insurance administration was based on A.M. Best data for private carriers and Social Security Administration data for state funds and the self-insured. Administrative costs of public insurance (mainly Medicaid and Medicare) amount to about 4% of the medical expenses paid by public insur-

Source: National Safety Council, *Accident Facts*, 1995.

ance, which were determined from Rice and MacKenzie (1989) and Hensler and others (1991).

Average police costs for motor-vehicle crashes were taken from Miller and others (1991) and multiplied by the Council's estimates of the number of fatal, injury, and property damage crashes.

Legal expenses include court costs, and plaintiff's and defendant's time and expenses. Hensler and others (1991) provided data on the proportion of injured persons who hire a lawyer, file a claim, and get compensation. Kakalik and Pace (1986) provided data on costs per case.

Fire losses were based on data published by the National Fire Protection Association in the *NFPA Journal.* The allocation into the classes was based on the property use for structure fires and other NFPA data for nonstructure fires.

Motor-vehicle damage costs were benchmarked to Blincoe and Faigin (1992) and multiplied by the Council's estimates of crash incidence.

Employer costs for work injuries is an estimate of the productivity costs incurred by employers. It assumes each fatality or permanent injury resulted in 4 person-months of disruption, serious injuries 1 person-month, and minor to moderate injuries 2 person-days. All injuries to nonworkers were assumed to involve 2 days of worker productivity loss. Average hourly earnings for supervisors and nonsupervisory workers were computed and then multiplied by the incidence and hours lost per case. Property damage and production delays (except motor-vehicle related) are not included in the estimates but can be substantial.

Lost quality of life is the difference between the value of a statistical fatality or statistical injury and the value of after-tax wages, fringe benefits, and household production. Be-

Source: National Safety Council, *Accident Facts,* 1995.

cause this does not represent real income not received or expenses incurred, it is not included in the total economic cost figure. If included, the resulting comprehensive costs can be used in cost-benefit analysis, because the total costs then represent the maximum amount society should spend to prevent a statistical death or injury.

Work deaths and injuries (p. 48). The method for estimating total work-related deaths and injuries is discussed above. The breakdown of deaths by industry division for the current year is obtained from the CFOI and state Accident Death Summary figures using the link-relative technique (also discussed above).

The estimate of nonfatal disabling injuries by industry division is made by multiplying the Council's estimate of employment for each industry division by the BLS estimate of the incidence rate of cases involving days away from work for each division (e.g., BLS, 1994) and then adjusting the results so that they add to the work-injury total previously established. The "private sector" average incidence rate is used for the government division, which is not covered in the BLS survey.

Source: National Safety Council, *Accident Facts*, 1995.

Source: National Safety Council, *Accident Facts*, 1995.